Praise for *States of Emergency*

"In this urgent new study, veteran international relations scholar Kees van der Pijl shows how the ruling groups exploited the Covid-19 virus to impose a worldwide state of emergency and a psychosis of fear that allowed them to extend their power and control at a time when global capitalism is beset by crisis. Brilliantly researched, impeccably sourced, the story is told in an engaging style and with great analytical acuity. Here is a dire warning against the slide into authoritarianism to contain the revolt of restive populations who can take no more deprivation. It must be read by the broadest possible public."

> – WILLIAM I. ROBINSON, Distinguished Professor of Sociology and Global and International Studies, University of California–Santa Barbara

"The Covid-19 event is fast shaping up to be a deception of epic proportions, whereby a constellation of powerful actors exploit events in order to usher through profound political and economic changes. More than ever, people are waking up to the fact that this is not about a virus... Derailing these nefarious agendas is without doubt the most urgent task facing humanity today and *States of Emergency* provides a powerful, thought provoking and critical starting point on the road to awareness and resistance."

> – DR. PIERS ROBINSON, Organisation for Propaganda Studies

"The Project for a New American Century wrote in 'Rebuilding America's Defenses' that 'advanced forms of biological warfare that can 'target' specific genotypes may transform biological warfare from the realm of terror to a politically useful tool.' They wrote it, they meant it, and we're now living it. Biological warfare, perfected at the cellular level, can now target distinct populations for annihilation under the guise of a 'naturally occurring' biological phenomenon. But, all populations are not equally subjected to the mortality of Covid-19. Who is going to die? Who constitute the surplus population no longer needed? How do we fight back when our bloodstreams are the battlefield? Kees van der Pijl goes further than most in addressing our current concerns and circumstances, but perhaps even that isn't far enough to capture the pure evil that confronts humanity today."

> – CYNTHIA McKINNEY, former Congresswoman and Green Party Candidate for President, most recently editor of *When China Sneezes: From the Coronavirus Lockdown to the Global Politico-Economic Crisis*

D0841553

STATES OF EMERGENCY

KEEPING THE GLOBAL POPULATION IN CHECK

KEES VAN DER PIJL

Clarity Press, Inc.

© 2022 Kees van der Pijl

ISBN: 978-1-949762-48-8
EBOOK ISBN: 978-1-949762-49-5

In-house editor: Diana G. Collier
Cover design: R. Jordan Santos

Cover photo, top: Seattle, Washington, USA—As protests grow, Seattle Police Department closes down roads surrounding the East Precinct on Capitol Hill. Photo by Derek Simeone

Cover photo, bottom: Los Angeles, California, USA, May 1, 2020— People in front of Los Angeles City Hall protest the state's COVID-19 stay-at-home orders in a "Fully Open California" protest. Photo by Matt Gush

Library of Congress Control Number: 2021947334

Clarity Press, Inc.
2625 Piedmont Rd. NE, Ste. 56
Atlanta, GA 30324, USA
https://www.claritypress.com

ACKNOWLEDGMENTS

This book originally appeared in Dutch in June 2021. I owe a debt to the publisher, Tom Zwitser of De Blauwe Tijger in Groningen, and to the editor, Dr Henk-Jan Prosman for his conscientious work on the text. I worked on an English version during the production process of that edition, adding additional information that had meanwhile become public, and was lucky to get the attention of Diana Collier at Clarity Press, for which I also thank Michel Chossudovsky of Global Research in Canada and British publishing legend, Roger van Zwanenburg.

In doing the research and writing, I owe a great debt to Karel van Wolferen, who like no other has used his experience as a foreign correspondent for Asia and academic to investigate the Covid crisis, on which he is meanwhile editing a successful biweekly, Gezond Verstand. Along with other print publications and Internet channels, this has become a major source of documentation without which I would not have been able to write the present study in a relatively short time.

Others who helped me in various ways with the current version include Gary Burn, Ab Gietelink, Elze van Hamelen, Olaf Harleman, Stan van Houcke, Alexander Kovryga, Bhabani Nayak, Örsan Şenalp, Jaap Soons, Sebe Vogel, and a number of anonymous Twitter correspondents. David Klooz kindly sent me his two self-published books on the subject. One of the leading figures in the democratic resistance in Germany, Ullrich Mies, just in time passed me two recent studies on the topic published in

Germany. Christine as always was my unfailing supporter in the undertaking. Diana Collier edited the present version with exceptional care and insight. Her erudition and profound understanding of the thrust and implications of the argument have greatly improved the original text.

Obviously none of these mentioned can be held responsible for the errors that remain.

TABLE OF CONTENTS

INTRODUCTION

S ociety as we know it—global capitalism with its home base in the West—has entered a revolutionary crisis. After years of preparation, the ruling oligarchy, which today exercises power across the globe, has seized on the outbreak of the SARS-CoV-2 virus and the respiratory disease attributed to it, Covid-19, to declare a global state of emergency in early 2020. This seizure of power is intended to prevent the Information Technology revolution (hereafter: IT revolution), the impact of which can be compared to that of the coming of the printing press at the end of the Middle Ages, from ushering in a democratic transformation.

In 2008 the capitalist speculation machine unleashed twenty years earlier came to a crashing halt. The casino was reopened after a short time with mainly water damage, or so it seemed; yet in the meantime, unprecedented unrest has arisen among the global population. Unlike the eve of World War I, the mass discontent this time has no clear-cut political orientation, as the IT revolution did not bring forth, as the Industrial Revolution had done, an organized revolutionary force such as the socialist labor movement, drawing its power from an industrial working class. With the decline of industrial production in the West, and concomitantly of unions, the unrest that arose after 2008 went in all directions—the Arab Spring, Occupy Wall Street, the Yellow Vests in France, and so on. Strikes, riots and anti-government demonstrations, as well as mass migration and drug abuse since then have broken all existing records—until the World Health Organization proclaimed the

1

Covid outbreak a pandemic. Governments across the globe swiftly followed by imposing states of emergency, which paradoxically
were then tightened as the virus subsided and henceforth followed
an obviously political calendar.

This book aims to address and dispel the psychosis of fear
into which the world has been plunged. In the process of research
and writing, I found that the 'pandemic' is not a simple, one-off
fraud, or a grand scheme cooked up by Klaus Schwab, the oracle of Davos, and obediently executed by national governments.
Rather, it is a complex, historical crisis, giving rise to a seizure
of power by the global ruling class that has been initiated from
different starting positions. Much about the Covid 'pandemic' is
still shrouded in mystery. It seems certain that the virus did escape
from a laboratory, but from which one, we do not know. What
we can conclude is that the official account of what is happening
around us is patently untrue and that it will therefore eventually
collapse. The timescale on which this will come about should not
be underestimated, as the mainstream media constitutes a key part
of the complex of forces that have seized power in this process; its
deception and propaganda regarding major historical events have
become routine since the 1990s.

What matters is that the Covid seizure of power, even more
comprehensively than previous states of emergency in the name
of terrorism, is working to prevent a democratic transition to a society beyond capitalism. The revolutionary crisis that has become
acute resides in the fact that governments have now taken their
populations hostage and cannot or dare not release them. This is
another reason why the entire effort at suppression is doomed to
end in failure. Too much has been set in motion too early, too
disjointedly, and the contradictions between the different interests
and institutions, only apparently all in agreement, are bound to
turn into overt conflict.

The book is organized as follows. In the first chapter, I begin
by presenting the key facts about the pandemic that make it
clear that what we are dealing with is not a medical, but a political

emergency. What is happening before our eyes is a step-by-step transition of Western liberalism to an authoritarian state and social structure, all in the name of 'the virus.' What effectively mirrors a state of war proclaimed in the spring of 2020, in reality is meant to safeguard the existing order. As George Orwell pointed out in his prophetic *1984,* all modern wars primarily serve that purpose. Yet in the West the state of emergency has antecedents different from those of the so-called contender states such as China. The latter's societies in a sense live under a permanent state of emergency, albeit with more positive returns, given the recognized achievement of China in lifting millions out of poverty and creating a burgeoning middle class. Hence, the way in which the population not only undergoes repression but responds to it differs too. In a country like China, people have been used to the limits of political involvement for generations. In the liberal tradition on the other hand, draconic measures are needed that can only be compared to psychological warfare and mental torture as the wealth gap becomes a chasm and social conditions tank.

Chapter 2 addresses the question of why this process was initiated if there is no real medical emergency. The comparison with the First and Second World Wars is pertinent here. Once again, there has been a rising tide of popular unrest bordering on insurrection in certain regions and countries. In the Middle East and in countries such as India, Chile and France, movements that were able to overthrow governments or had actually done so, had burst onto the scene, instilling fear in the ruling classes the world over. With the Covid state of emergency, the popular movement, in all its diversity, has been frozen for the time being. The specific social structure of North America, Australasia, and Europe so far has worked to facilitate this quasi-normalization. On the one hand we have a cosmopolitan cadre that works for the oligarchy and is concentrated in the big cities. It shares the urban space with a growing immigrant population mainly there to serve it. Facing it is a marginalized domestic population that has become largely redundant. In this complex configuration of forces a political stalemate has crystallized in which the labels 'left' and 'right' are losing their cogency, but which retains a revolutionary potential, nevertheless. The chapter also details the shadow structures of repression that had accompanied the prior era of class compromise. These have now come out into the open, as governments adopt counterinsurgency methods to deal with the growing resistance to the Covid emergency shutdowns.

Chapter 3 analyzes the Covid crisis as a seizure of power to force a 'new normal' on society. All exercise of power in liberal capitalism rests on a social contract bolstered by accompanying ideology, a comprehensive concept of control that substitutes for the aggregating role previously played by religion, the nation or civilization. This time the ruling class has chosen not to wait for a 'new normal' to arise organically from the process of class formation, as occurred after World War II and again in the 70s. Capitalism can no longer engender a rational class compromise and instead has begun to rule through worst case scenarios. The new power bloc that emerged from the intelligence needs of the U.S. national security state, much of it privatized as IT monopolies and sprawling (multi-) media conglomerates, has imposed the Covid scenario from above by an external shock meant to create a surveillance society. Finance had profited from IT innovations but after the 2008 crash, the riskiest forms of speculation were reined in by restructuring and aggregating financial control in the form of 'passive index funds' such as BlackRock. Whether the Covid crisis was seized on to forestall an impending financial collapse, or to prevent the populist U.S. president, Donald Trump, from being reelected, or both, cannot be determined with certainty. Nationalist populism, which seeks to overcome the political crisis of Western democracy by mobilizing discontent against the privileged urban cadre and against immigrants presents itself as a revolutionary force, much like fascist movements did in the 1930s. Yet in current conditions, the mainstream in the oligarchy does not seem to need this diversionary force for the time being.

Chapter 4 demonstrates that a pandemic, real or imagined, has become an ideal cover for establishing a surveillance society without resorting to overt violence. After the fear of terrorism eroded, the presumed threat of 'Putin,' the specter of climate change, and other worst-case scenarios proved unable to mobilize society to the same degree. The outbreak of an unknown infectious disease instead has proven to be a highly effective new installment of the politics of fear on which government legitimacy in the West relies after the disintegration of class compromise and the collapse of Soviet state socialism. At the turn of the century, SARS-1, bird flu and, in the aftermath of the financial collapse, the Mexican or swine flu panic of 2009 showed what was politically possible with a virus alert even though those epidemics were not widespread enough to allow the imposition of a state of emergency. Yet the evaluation of the lockdowns in China and Canada at the time of SARS-1 showed that citizens are willing to undergo such a radical intervention as a test of their citizenship, even of their patriotism. In 2010 the Rockefeller Foundation came up with a detailed scenario for an imaginary pandemic that would permit mass repression. In the years that followed, the script for an integral shutdown of society was worked out in detail. Here the Gates Foundation of Microsoft founder Bill Gates, the exponent par excellence of the IT power bloc, served as the switchboard by which the virus scenario was passed on to the WHO, national governments, and the actual biopolitical complex, discussed more extensively in Chapter 6.

One of the most uncertain factors in the Covid crisis is the relationship between the West, and the United States in particular, and China. Chapter 5 shows that the U.S. has established a comprehensive biological warfare research and development infrastructure targeting Russia and China, with black Africa as an additional testing ground. But paradoxically, the U.S. also was in close cooperation with China in the field of microbiological research, even though China is a contender for power which, in the IT domain for instance, was treated as an adversary. In the course of 2019, U.S.-Chinese cooperation on biodefense too went awry, with Canada involved as well. That 'the virus' has escaped from a laboratory where viruses undergo gain-of-function to make them more dangerous seems certain, but whether this was the laboratory

in Wuhan to which U.S. research had been subcontracted, or Fort
Detrick in Maryland, is uncertain. The chapter concludes with the
observation that despite the transformation of the liberal West to
the authoritarian Chinese model (though undoubtedly still controlled by its present elites and geared to protecting their interests),
it is unlikely, in light of the rapid shifts in the balance of power,
that this will also lead to a stable, 'ultra-imperialist' truce in their
mutual relations.

Chapter 6 discusses the pandemic as disaster capitalism,
Naomi Klein's term for the economic opportunities for business
created by major catastrophes. In this case, the opportunities are
for the biopolitical complex, especially the pharmaceutical industry, the biotech sector, the Gates Foundation, and medical schools
and research centers such as that at Johns Hopkins University. The
internationalization of state policy, with individual governments
implementing guidelines set at the global level, provided the
channels through which the intelligence-IT-media bloc, joining
forces with the biopolitical complex, were able to push through
the imposition of the Covid state of emergency. Throughout the
course of 2019, we see how a series of planning meetings not only
prepared for a possible virus outbreak but in particular focused on
the 'infodemic' of dissenting opinions, highlighting the political
thrust of the presumed medical emergency.

The closing Chapter 7 examines the possibilities of the IT
revolution for a different course, one aimed at radical democracy
and digital planning. What is special about the IT revolution is that,
for the first time in history, the contradiction between individual
freedom and collective social and ecological security has the technology allowing it, in principle, to be overcome. The ruling class
of the capitalist West is aware of this potential and wants to nip it in
the bud, beginning with the massive clampdown on social media.
The ruling classes of non-Western countries are also interested
in IT restriction and surveillance, if they have not already taken
important steps themselves towards instituting it. While planning,
like (state) socialism, today suffers from a bad reputation after
the stagnation and collapse of the Soviet command economy, all

major corporations use digital logistical systems. This chapter addresses how, in Soviet times, remarkable initiatives were taken to implement such a digital planning system on a national scale, although due to bureaucratic conservatism and lack of democracy this ran aground. Again in Chile, a comparable experiment was cut short by the Pinochet coup. This time it is different. The world has been forced into a revolutionary situation by the oligarchy and now faces the choice to submit or opt for a viable alternative that would entail the dispossession of the billionaire owners of what Marx called 'the social brain.' In the process a broad, politically heterogeneous movement for freedom will emerge that will restore and renew democracy whilst exploiting the possibilities of the IT revolution for a viable human future—or perish.

1.

THE COVID CRISIS
AS A STATE OF SIEGE

Following the World Health Organization (WHO) declaration
that the respiratory infection caused by the corona virus
SARS-CoV-2 (hereafter, SARS-2) was a *pandemic* on 11 March
2020, a state of emergency was introduced practically across the
globe. It is perceived by many in our part of the world as the most
serious breach of fundamental rights and civil liberties they have
ever experienced, comparable only to a state of war. Everything
now indicates that if it were up to those in power, the state of
emergency would acquire a permanent character. When mortality
rates fell after a few months, governments switched to counting
cases, now termed 'infections,' on the basis of positive PCR tests,
which are expressly *not* suited for that purpose. Thus, the mass
psychosis induced in the spring of 2020 was maintained.

This chapter begins with a brief overview of the grounds on
which it can be established with a high degree of certainty that the
Covid crisis is a fraud, serving as cover for a political seizure of
power. I then discuss what is the equivalent of a state of war being
imposed on the global population and its implications. For that
purpose, I distinguish between three types of a state of emergency,
which have merged into each other in the current situation: the
original state of emergency in liberal democracies; the permanent
one existing in so-called contender states opposing the hegemonic West; and the global state of emergency which the French

philosopher, Michel Foucault, derived from what he labeled *bio-politics* in the 1970s. We conclude with a comparison between the current psychological warfare offensive against the population, aimed at paralyzing all forms of resistance, and mental torture.

A RESPIRATORY DISEASE OR A POLITICAL AGENDA?

Several months of intensive consultations had preceded the WHO's declaration of the pandemic on 11 March. Although suspicious cases of a similar illness had been evident in the preceding period in other countries, the first three cases of Covid-19 were officially reported in the city of Wuhan on 27 December 2019, two weeks before some 400 million Chinese were about to travel to their families to celebrate the Chinese New Year. On 23 January, two hours after midnight, the authorities in Wuhan announced that at 10:00 the next morning a complete lockdown would enter into force for the city of 11 million. By that time, hundreds of thousands, if not several millions of residents had already fled in panic to avoid the quarantine. Meanwhile those gathered for the World Economic Forum in Davos, Switzerland, from 20 to 24 January, in the presence of the Director-General of the WHO, the Ethiopian, Dr. Tedros Adhanom Ghebreyesus, discussed how to respond to this event. Tedros flew to China afterwards to discuss matters with President Xi Jinping on 28 January. Two days later, the WHO declared a 'Public Health Emergency of International Concern,' whilst its chief praised the Chinese lockdown as an unprecedented but decisive step. Another five weeks later, the world health organization officially declared a pandemic; at the time, according to WHO, there were an estimated 118,000 cases of Covid-19 in 114 countries, with 4,291 recorded deaths.[1]

Meanwhile, images of the Chinese metropolis that looked like it had been abandoned by its last inhabitant, had been circulated by the media around the world. 'Wuhan was an unforgettable spectacle that had a major impact on the Western psyche,' Patrick Henningsen commented. 'When the corona virus hit the European and North American shores, the public was already conditioned to expect a Chinese-style response from their own governments. Not surprisingly, this was exactly what they got and in fact it was what they *demanded.*[2] Amid apocalyptic expert predictions in the media that we were facing an unprecedented disaster that could only be averted by the most draconic means, governments fell over each other to heed the call to arms and introduce drastic measures.

The media, especially the talk shows on TV, assumed the role of propaganda channels right from the start, excluding any dissenting voices. Parliaments too fell in line without a murmur. However, when the peak passed and the 'pandemic' turned out to be much less deadly than predicted, the restrictive measures remained in force, despite the tremendous damage they had already caused economically and socially, especially to mental health. At this writing, lockdowns and the easing of restrictions seem likely to follow a wack-a-mole pattern, as the media resound with announcements of new and even more threatening variants to back up the call for vax-passes in one country after another.

Initially, the lockdowns were justified by reference to limited hospital capacity, without mentioning that decades of neoliberal austerity had reduced the ability to handle a health emergency. In China, new hospitals were being erected at breakneck speed, but in the rest of the world steps to upgrade the battered medical infrastructure were few and far between. There were a few countries which refused to heed the imposition of the global state of emergency, such as Tanzania or Belarus, which would suffer no substantially different rates of infection and mortality compared to their neighbors. Belarus was offered $940 million by the World Bank if it imposed lockdowns and a curfew but declined the offer.[3] The president of Tanzania, John Magufuli, who had ridiculed the epidemic by having plant and animal tissue tested 'for Covid' and with several samples found positive, died in early 2021 at the age of 61, as did his Burundi counterpart, who also had declared the virus to be nothing extraordinary.[4]

Within the EU, Sweden followed a course that was clearly different from the rest. According to David Klooz, a Canadian health officer with more than 30 years of experience, it was one of the few countries to have a fact-based Covid policy. In mid-June 2020, after the peak in hospital admissions and mortality in both Sweden and the locked-down Netherlands had passed, 6,057 Dutch people had been registered as Covid deaths, 0.035 percent of the population; in Sweden, 4,866; 0.048 percent. 9.6 percent of the Swedish cases died; in The Netherlands, 12.5 percent. Across the age groups, the picture was practically the same: one-third of the deceased were over 90, one-third were between 80 and 89, one-fifth to a quarter in their seventies, and 6 to 8 percent in their sixties.[5] Although Covid-19 is not a flu (the origin of the virus in a bio-warfare laboratory will be discussed later), it has a similar casualty profile in terms of age groups: mainly older people with poor health and/or underlying diseases. As early as March 2020, in relatively hard-hit Italy, for example, half of the deaths attributed to the virus suffered from *three* other serious illnesses; one-quarter from two, and the remaining quarter from one. Only 0.8 percent of those who died appeared in good health.[6]

The WHO initially assumed a mortality factor (the percentage of people infected and dying from it) of 3.4 percent, but at the beginning of October 2020 this was scaled down to 0.14 percent; the deaths attributed to Covid had then just passed one million worldwide (in the great flu wave of 1968–69, there were one and a half million victims), while the number of infections was estimated at one-tenth of the world's population. Even Klaus Schwab, founder and animator of the World Economic Forum and author of *The Great Reset*, a blueprint for the 'new normal' that should take shape after the 'pandemic,' wrote in that book that so far (July 2020) Covid is 'one of the least deadly pandemics the world has experience[d] over the last 2000 years' (by comparison, HIV-AIDS has killed some 30 million people since the outbreak in the early 1980s).[7] At the beginning of June it had in fact become clear that the SARS-2 virus had passed its peak. The head of the San Raffaele hospital in Milan, in the initially hard-hit region of Lombardy, stated that the virus *no longer existed in a clinical sense* and that his country should return to normal conditions as soon as possible. The former director of the immunology institute at the University of Bern, Stadler, said the same. The prime minister of Norway, Erna Solberg, in late May 2020 even apologized on TV for the lockdown, which according to her had only been introduced on the basis of 'worst case scenarios' wrongly taken for granted.[8]

So why continue with the socially and economically disruptive 'measures'?

THE SWITCH TO TESTING AND 'INFECTIONS'

The answer to this came after the summer, when death rates had fallen further. Now, government health institutes such as the Centers for Disease Control (CDC) in the United States, the Robert Koch Institute (RKI) in Germany, and their equivalents in other countries, switched to counting people 'tested positive' for SARS-2. These were then referred to as 'infections' with the suggestion that, even when no symptoms were in evidence, they themselves could cause new infections. In order to maintain the impression of rising numbers even as the 'pandemic' subsided, the figures were consistently published without mentioning the number of tests administered or any other reference to put them in context.

The test in question, PCR (*Polymerase Chain Reaction*), is intended to multiply molecules for microbiological research. After amplifying the sample 30 times, the researcher already has a billion molecules, in which traces of genetic material may indicate, in this case, a (residual) corona virus. Adding more cycles makes the test meaningless. The inventor of the test, Dr. Kary Mullis, who won the Nobel Prize in Chemistry for it in 1993, stated several times before his death in 2019 that a PCR test of viral RNA residues (nucleic acid in a single helix such as SARS-2) cannot be used for a medical diagnosis; what will be found depends entirely on the choice of the so-called *primer*.[9] In April 2020, *The Lancet* published a letter pointing out that viral RNA can be detected long after an infection. The immune system may deal with a virus in different ways, but afterwards allows the viral RNA to exist for weeks, for example 6 to 8 weeks in measles.[10] In short, a positive PCR test does not mean contamination, let alone infectivity.

Nevertheless, the German virologist, Prof. Christian Drosten, published a PCR test for SARS-2 (then still largely unknown) in the journal *Eurosurveillance* in January 2020. Drosten had already described such a test for SARS-1 in 2003 and successfully marketed it through a Hamburg biotechnology company. In the same year he was awarded a doctorate and although his doctoral thesis remained untraceable until 2020 and he became a professor without the usual exam (*Habilitation*), his fame grew, and he was finally appointed chief virologist at the Berlin Charité hospital.[11] The article in *Eurosurveillance* (the organ in which the spread of infectious diseases is reported so that prevention and control measures may be taken) was co-signed by authors from nine European countries. It turned out to have been accepted without peer review and to contain a series of errors. The Dutch microbiologist, Dr. Pieter Borger, wrote to the authors, to the journal, and to the responsible EU institution, and when no satisfactory response was forthcoming, submitted a detailed request for retraction with 10 other experts. This was duly subjected to peer review and rejected in January 2021. Borger had raised his criticism already in the spring, but it fell on deaf ears in the media, and he was eventually also removed from LinkedIn.[12]

Meanwhile the PCR test was in use everywhere for Covid detection. The test requires three genes for evidence of a virus, but from September, in several countries the number of cycles was raised to 35 and above, and the required number of genes reduced to two or even one, making the results meaningless. Now the alarm was sounded about a 'second wave,' no longer based on deaths or illnesses, but on rising numbers of positive tests ('infections'). In July 2020, even the *New York Times* conceded that 85 to 90 percent of those tested positive that month on the basis of 40 PCR cycles would have been negative for 30.[13] Why it took the WHO until January 2021 to admonish users of the PCR test to follow the directions for use and consider the test as complementary to actual diagnosis, must be left for insiders to explain.[14]

In the meantime, measures spread across the globe in the name of the supposedly raging 'pandemic,' whilst authoritarian controls proliferated, and basic rights were suspended one after another. Closure of schools and universities, obligatory wearing of face masks, social distancing, even curfews and in some countries, a prohibition to leave one's home, were being introduced. The ensuing social disaster, of which elderly residents of care homes, children, young people, young parents, and of course, entrepreneurs including self-employed workers, are the primary victims, was not based on a medical emergency. It results from manipulation and an information war, which the authorities are waging against their populations. This war is covered with unflinching support by the mainstream media, whilst Internet platforms are applying censorship to remove all possible information about the availability of an effective existing medication (*Hydroxychrloroquine* or *Ivermectin*) and about the dangers of vaccination with hastily developed gene therapies, to all of which I come back in Chapter 6.

On the basis of the foregoing, we may conclude that there is no proportionality whatsoever between the actual medical emergency and measures that in Europe recall the Nazi occupation, in the United States, the years of McCarthyism, and so on. The Covid condition, certainly after the summer of 2020, appears to be eminently manageable in every way. Nonetheless, politicians and media claim that there continues to be an exceptional danger. So something else is going on.

THE STATE OF WAR IN PEACETIME

One of the many relevant insights in George Orwell's novel *1984* is that in contemporary society, war is no longer primarily directed against a foreign enemy. The best-known aspect of this rightly world-famous book is the permanent surveillance via TV screens (*'Big Brother is watching you'*). Yet the protagonist of *1984*, Winston Smith, still manages to read a scientific study criticizing the society he lives in. This fictitious study is by one Emmanuel Goldstein, the alter ego of the Russian revolutionary, Leon Trotsky (real name, Bronstein). Like the earlier *Animal Farm, 1984* was inspired by the corruption of the Russian Revolution under Stalin's dictatorship. Trotsky was eventually murdered in exile in Mexico by a Soviet agent. In *1984* he is the model of the Enemy of the People, the target of orchestrated popular anger in the daily 'Two Minutes Hatred,' a bit like that in the West directed against 'Putin' in our time.

War, Smith reads in Bronstein/Goldstein's fictional *Theory and Practice of Oligarchical Collectivism,* is pure deception in the present day. 'It is now a purely internal matter.' And the text continues:

> In the past, the ruling groups of all countries, although
> they might recognize their common interest and
> therefore limit the destructiveness of war, did fight
> against one another, and the victor always plundered
> the vanquished. In our own day they are not fighting
> against one another at all. *The war is waged by each
> ruling group against its own subjects, and the object
> of the war is not to make or prevent conquests of
> territory, but to keep the structure of society intact.* The
> very word "war" therefore, has become misleading. It
> would probably be accurate to say that by becoming
> continuous war has ceased to exist.[15]

It is often overlooked that the First and Second World Wars were also to a large extent counterrevolutionary episodes in this sense. The First World War worked to interrupt a process of ongoing democratization: the German ruling class, having noticed events in Russia in 1905, feared the socialist workers' movement in spite of Bismarck's prior success in nationalizing it to a large degree; the British rulers feared an uprising in Ireland, the great

European empires feared the nationalities within their borders, etc. The Russian Revolution did occur after three years of bloodshed, but the democratic political culture was irreparably damaged. Fascism and National Socialism, but also Stalinism in the Soviet Union, cannot be understood without the bloodbath of 1914–18. The Second World War turned into an even more cruel, ruthless massacre. Once again, the need to 'keep the structure of society intact' was at the forefront, although once again even greater social explosions ensued after the end of hostilities.[16]

The socialism that threatened the ruling classes of the first half of the twentieth century grew out of the Industrial Revolution and the war against it was mechanical, aimed at the physical extermination of the masses, once overseas settlement no longer provided an escape hatch due to mounting rivalries. Today we live in the age of the IT revolution, and both the democratic challenge and the answer to it have moved into the realm of *information*, the prevailing thought patterns and the general state of mind of the subjects. In the West, especially since the collapse of the Soviet bloc, the authority of governments no longer rests on a real social contract but on a politics of fear, and following the attacks of 11 September 2001 ('9/11'), that too has become permanent and the transition to 'a war waged by every ruling group against its own subjects' irreversible. The revelations by Edward Snowden about the global eavesdropping practices by the Five Eyes, the signal intelligence agencies of the English-speaking West led by the National Security Agency of the U.S., have also revealed that there is no longer any doubt either about *1984*'s other key message: *Big Brother Is Watching You.*

That the war serves other purposes than 'victory' was underlined in an article in the *Washington Post* on the occasion of the 10th anniversary of 9/11. In a report from Fort Campbell Army Base in Kentucky, the writer explains that the idea that a war at some point is over no longer applies. In the past, the military and the general public saw war as an exception; peace was the norm. But now, according to a Pentagon report, we have entered a period of permanent conflict, one that the developed world should not have the illusion of winning. People, according to the writer, must learn to live with a degree of insecurity and fear.[17]

We are now an additional ten years later. 9/11, the Patriot Act, the proclamation of the War on Terror, and the publication by WikiLeaks of the crimes committed by the U.S. and Britain in that war, were followed by whipping up the Russian threat and the alleged contamination of our news supply by 'Moscow.' A never-ending stream of largely fabricated threats is causing fear. Despite this, the financial collapse of 2008, as we will see, was followed by an unprecedented resurgence of social unrest, from strikes to riots, even full-blown uprisings—all of which required a fresh response, for which the IT sector and the billionaire oligarchy behind it had the resources ready.

THREE TIERS OF THE STATE OF EMERGENCY IN WORLD CAPITALISM

The permanent war situation described in *1984* is not only a safety valve to keep the population at home under control by directing pent-up domestic unrest that threatened the system toward actual war elsewhere. In *1984* it is nominally waged between three continents, namely Oceania, Eurasia and East Asia. 'Oceania' refers to the English-speaking world, viz., the British Isles, North America, Australia and New Zealand and white South Africa. Winston Smith lives in England, Airstrip 1 of Oceania, right across from Eurasia, with which it is 'at war.'

The real Oceania was the most successful of the new, overseas empires that were established after the voyages of discovery from Europe at the end of the Middle Ages (those of Spain, Portugal and the Netherlands). It was with the first English-speaking colonies in North America in the 17th century, that the specific balance was reached between central state authority and the self-regulation of the property-owning classes. Only under these conditions was capitalism able to develop as free enterprise, and yet be fostered and protected. The other three European colonial states essentially remained trading powers, their states ill-equipped for the purpose, and if they transplanted civilians to their colonies, these mostly mixed with the indigenous peoples. The English-speaking peoples, on the other hand, pursued a policy of ethnic cleansing against the native inhabitants and also strictly segregated the enslaved African population they brought to the colonies in the Western Hemisphere.

The original liberalism relied on freedom and civil rights, but always had a built-in emergency brake in case these cannot be maintained by law-based procedure. In his *Two Treatises of Government* of 1689, John Locke outlined this particular state/society configuration in detail. Locke was the ideologue of the free traders and commercially minded landowners who came to power a year earlier thanks to a regime change operation by the Dutch stadtholder, William of Orange, who was then made non-hereditary king of England. In Locke's model, the property-owning class is sovereign (in the sense of the highest authority); the state (or the king) enjoys a *derived* sovereignty. The emergency brake was available in case property rights were threatened and would suspend the law altogether. To this end, the rulers enjoy a 'prerogative,' according to Locke: it entitles them to apply power at will—'*doing publick good without a Rule.*'[18]

Liberal capitalism, then, is always protected by a state of emergency kept in reserve; otherwise, there is no need for a strong state permanently imposing itself on society. Towards the outside world, capitalist liberalism was expansive and extremely violent; internally, it was nominally rule-based but the rules in English-speaking liberal society were applied harshly and still are today.[19] Yet the balance between central authority and a self-regulating society was not established once and for all in Locke's days. It had to be fine-tuned twice after the English Civil War, essentially an *Atlantic civil war*—first, the American War of Independence, and then, in the American Civil War—before establishing its global hegemony. The triumph of the capitalist West in the 19th and 20th centuries as a liberal heartland can only be understood as a result of this repeated coordination of state authority and liberalism, not of liberalism alone.[20]

The second tier of a state of emergency originates in the states which from the 17th century onwards, in order to defend themselves against a commercially superior England, developed political formations in which *state power takes precedence over all social initiative*. Any other course would make such a country defenseless, open to colonization or otherwise subjection to the expanding Anglophone West. In these *contender states*, of which France was the prototype at the time of Louis XIV and again after the Revolution of 1789, under Napoleon, a different state/society relation obtains. It is not Locke's scheme that is the guiding principle, but (we may infer) that of Thomas Hobbes, who wrote at an earlier stage of the English Civil War. In his *Leviathan* of 1651 Hobbes suggested that a society cannot possibly be left to self-regulation; that would unleash a war of all against all. Hence state power must intervene *permanently* in every aspect of social life. At the time Hobbes wrote, the concept of *coup d'état* was developed in France to capture the shock-like interventions by which central authority extends its power over society and positions itself as the systemic guide of its functioning in every respect.[21]

All subsequent contender states (unified Germany, Japan after the Meiji Revolution, the Soviet Union, and China, to name only the most important) would adopt this position against the liberal, English-speaking West. They have maintained a more or less *permanent state of emergency*, typically confirmed by 'coups,' revolutions from above (sometimes following popular revolutions).

We see therefore that in this type of state/society complex, the state of emergency is institutionalized differently from that in the liberal West. Not only is liberalism in the sense of individual self-determination and self-regulation under the law, weak or absent, but capitalism as an economic system is also subject to state supervision or even, as in the USSR, formally abolished. The rulers in such contender formations (the 'state class') have other priorities than protecting private property. That was evident when the Second German Empire, after the defeat in the First World

War, briefly mutated to the Weimar Republic, which had to do *without* a tentacular state and a permanent state of emergency. To us, this episode is again of great significance because the state of emergency today established in Europe and many other places is more similar to the German experience in the 1920s and 1930s than to Locke's original scheme.

The state of emergency in Germany, which ultimately led to Hitler's appointment as Chancellor, was the central reference for the ultra-conservative legal scholar, Carl Schmitt. In 1922 Schmitt defined the *sovereign* (that is, the supreme power in the country) as the one who decides that a state of emergency exists.[22] Yet, even after the state of emergency had been instituted in July 1930, Schmitt warned that in a situation where none of the opposing classes prevailed, the constitution itself (after all, the Weimar Republic continued to exist) becomes unstable. In such an existential crisis, he wrote in 1931, the president of the republic, endowed with a mandate from the people, must 'preserve the unity of the people as a political whole against the multitude of social and economic power groups and *exercise the political will directly.*'[23]

This then happened when the president, Marshal Hindenburg, installed Adolf Hitler as Chancellor in January 1933. Within a month the Nazis had consolidated their power with the help of the Reichstag fire and the following year the plebeian SA leadership, among whose followers the idea of a revolution against capital (the 'Jewish plutocracy') was still alive, was massacred by the SS in the 'Night of the Long Knives.' Schmitt then produced a short text in which he radicalized the reasoning of his 1931 book to account for the new situation, claiming that the law has its origins in the institution of the *Fuehrer.* 'In the supreme emergency the highest law is confirmed ... all law has its origin in the right to life of a people.'[24] This 'right to life of a people' as a ground for the suspension of democracy has an eerie resemblance to the role of the health motif in the current corona dictatorship.

Thus, the state of emergency resulted in a return to the contender constellation, in this case in the form of fascist dictatorship and war. After the defeat of the Axis Powers, the USSR and the Soviet bloc became the new contenders. The losers lost their sovereignty to the victors, 'Oceania,' and after the collapse of state socialism, the same would have happened to Russia and China if these countries had not been pushed back into the contender role in spite of their conversion to (state-oligarchic) capitalism. Nevertheless, even these states to a significant extent have submitted to global governance, particularly through their membership in the United Nations and its functional organizations: UNESCO, UNICEF, the Food and Agriculture Organization (FAO), the International Labor Organization (ILO), the World Trade Organization (WTO), and most importantly for the present discussion, the World Health Organization WHO.

In the Covid crisis, China and, to a lesser extent, Russia, have thus also become engulfed by what Robert Cox calls the 'internationalization of the state,' in which sovereignty is wholly or partly shifted to the international level. From that level the overall direction of policy, sometimes even the details, are handed down to national governments. In each country, depending on the possibilities in a given balance of power, these are then carried out. In each, the power of the international ruling class is separately applied in a *concentrated fashion*.[25] This is clearly visible in the current crisis.

The state of emergency too in these circumstances gravitates to the international level. According to Foucault, 'political rule is [then] practised through a complex triangle of "sovereignty, discipline and governmental management, which has [the] population as its main target and [uses] apparatuses of security as its essential mechanism."' Foucault denotes this with the term *biopolitics*. The biopolitical state of emergency on the one hand transcends the level of the individual state, but power also reaches deeper and *'takes charge of life itself.'*[26]

As often with Foucault's analyses, it is left open *who* wields power, but the idea itself nonetheless remains crucial. The Covid crisis, in which 178 countries are fulfilling their earlier treaty obligations under the WHO to take a series of prescribed measures

once a pandemic is declared, seems to fit perfectly with his definition of a global state of emergency. We have the remainder of this
book to find out who are the driving forces behind this intervention and why they resorted to it.

VIOLENT TRANSITION TO A 'NEW NORMAL'

U nder the Covid state of emergency, a biopolitical program is
carried out, viz., a political project that wants to connect the
population directly, possibly even physically, to IT networks. The
peak of the 'pandemic' may have passed by the summer of 2020,
but the state of emergency must be preserved, otherwise the need
for 'vaccination' (in reality, an experimental gene therapy) cannot
continue to be justified much longer. The required level of fear is
being maintained by 'rising numbers,' at first by the daily reports
on the number of deaths, dramatized by stories of mass graves and
predictions of millions of victims; before switching from 'cases'
to 'infections,' supposedly established by PCR tests, and then by
repeated alarms about new variants of the virus accompanied by
dire predictions of what might happen, and so on. This is the first
pandemic in history that is not self-evident to everybody but rather requires mass indoctrination on the basis of extensive 'testing,'
media censorship, etc. If it were not for same, we would not know
we are in the grip of a health emergency.

The strategy to keep the population in a state of fear can
only be reconstructed from documents that have emerged in a few
countries. Thus, an advisory group set up in Britain during the
Mexican or swine flu epidemic in 2009, the *Scientific Pandemic
Influenza group on Behaviour and Communications* (SPI-B & C),
was reactivated in the Covid crisis. Its behavioral branch, SPI-B,
immediately adopted an aggressive approach in the face of the
new health threat. It established that people felt insufficiently
threatened at the personal level and therefore recommended
steps to raise the level of the sense of danger. Hard-hitting, in-
tensely emotional messages were deemed most suitable, to be

accompanied by steps to mobilize 'social disapproval' of those ignoring or resisting sanitary measures, lockdown, and the like.[27]

The SPI-B conceded that it was necessary to guard against dangerous situations arising from certain people being made scapegoats. There had already occurred aggression against people of Chinese descent, who were held responsible for 'the release of the virus.' Nevertheless, a strong collective identity should be promoted, the SPI-B insisted, by calling on the public's sense of responsibility for others, solidarity and the like (sentiments rarely or never invoked in matters of social inequality, homelessness, foreign wars, and so on). The government, it advised, must achieve this by a targeted approach, setting apart and isolating critics, whilst the media should keep the public in a state of tension by alarmist reporting.

Far more detailed documents on how to maximize fear among the public emerged in Germany, which plays a leading role in Europe's approach to the Covid state of emergency. According to leaked e-mails between the German Interior Ministry and various research institutions and universities immediately after the proclamation of the pandemic, the Federal Government demanded at short notice computer models and other scientific tools in order to impose 'preventive and repressive measures in the short term.' The Interior Minister, Horst Seehofer, and his state secretary, Markus Kerber, demanded that a *worst case* scenario be worked out that could legitimize an extreme lockdown. Seehofer took action after a visit at his home by the aforementioned virologist, Prof. Christian Drosten and Dr. Lothar Wieler, head of the RKI health institute. On 19 March, Kerber urged the RKI and the most important professional bodies in the field of economics and political science, as well as individual universities, to provide the required 'scientific' basis for a drastic clampdown within four days. Instead of detailing the medical situation, he explained to the addressees that this demand was a matter of *'internal security and the stability of public order.'*[28]

The advice requested by Seehofer and his associates was incorporated into a secret document on how to induce fear and docility. The 'feeling of powerlessness must be curbed by the impression of strong state intervention,' ran one of the expert opinions. One scenario outlined that the public should be conditioned by images of 'many seriously ill being taken to hospital by relatives, only to be rejected and upon return home, dying helplessly whilst gasping for air,' etc.[29]

The willingness of academia to submit such extreme disaster scenarios by return mail without first inquiring what was actually happening, is painful evidence of the corruption of contemporary academia. In the end the Federal government left it to a scenario according to which there would be one million Covid deaths, thus justifying the lockdown.

PSYCHOLOGICAL WARFARE AND TORTURE

The state of emergency that has been declared is fundamentally sustained by psychological warfare. According to the Italian judge, Dr. Angelo Giorgiani, we are dealing with a new form of terror on the part of the state, in which he distinguishes three phases.[30] The first phase consists of eliciting fear and uncertainty. This is achieved by the aforementioned examples of intentionally dramatizing the situation; by denigrating and not using available medication (or even outlawing it) and by-passing frontline doctors; by taking away basic freedoms from people who are not ill, and by locking down the economy.

The second phase consists of a 'messianic' announcement that vaccines are on the way and that relaxation of the lockdowns will become possible once they have been administered on a sufficient scale. This raises hope that the absurd situation in which people find themselves will be lifted soon *if only the vaccines arrive in the required quantity*. Questions such as the evolution of the virus, the actual seriousness of the infection, the chances of a fatal outcome, and the like, all recede into the background; if only the vaccines arrive, our lives can return to normal! The third phase consists of the actual inoculation campaign with the revolutionary gene therapies. At the time Giorgiani gave his interview, only Israel had an advanced program up and running to force the population into accepting these 'vaccines.' These are being administered on the basis of an agreement with pharmaceutical giant Pfizer that Israel will provide the company with the percentages of the population to which the gene therapy has been applied, which side effects have occurred, etc.

Again, frontline doctors are being by-passed as the state imposes draconic measures to keep the campaign going, excluding people from access to all kinds of amenities and public services if not 'vaccinated.' Meanwhile, Britain, Portugal and the United Arab Emirates too have gone down the path of mass vaccination. We will return to this in Chapter 6.

That an Italian judge could speak of a new form of state terror should be understood in light of the 1970s Strategy of Tension in that country where successive governments and state institutions such as the intelligence services engaged in targeted assassinations, false flag bomb attacks and other acts of terror, in order to block the Communist Party from government participation.[31] Giorgiani himself served as an anti-mafia judge, another social sector with which parts of the state colluded. So, whereas in a country like The Netherlands, a health minister declaring the contracts with vaccine producers a *state secret*, is broadly believed to have valid reasons for doing so (certainly by parliament and the media), the Italian experience has left Italians with a much deeper distrust of the state and its officers. Other countries whose experience with state authorities systematically acting in bad faith or worse predisposes them towards much greater resistance to the current state of emergency, according to Giorgiani are Uruguay and Poland, and there will be others.[32]

In most countries, however, the bulk of the population so far has remained largely *passive* in this upheaval and submits to the revolution from above, intended to implement a new, oligarchic IT capitalism to which we will return, later. In the novel *Il Gattopardo* about Italian unification, a key example of such a passive revolution, this is summed up with the famous line, 'If we want things to stay the same, they will have to change.'[33] But then, as in other top-down revolutions, the population was given a subsidiary role (in Italy, the Garibaldi movement). In the Covid crisis, the lockdown measure has the capacity to prohibit *any* political activity, insofar as freedom of assembly and demonstration, having once been abolished, can be re-abolished as new variants in the virus emerge. Moreover, the sovereign right to control over one's body, its *habeas corpus*, is being suspended to allow vaccination to be made compulsory. Chancellor Merkel's denunciation of criticism of the lockdown measures as 'an attack on our way of life' may sound grotesque, but it indicates that the governing class and the ruling class for whom they play their part, recognize that the Covid state of emergency serves to save their particular social order.[34]

In her 2007 book *The Shock Doctrine,* Canadian journalist Naomi Klein describes how two changes occur in an existential crisis: the past is erased and a 'new normal' takes its place. That was the case for Chile and Argentina in the 1970s, and for the Soviet Union when it collapsed in 1991.[35] The same is now true of the Covid 'pandemic.' This brings us to the comparison with *torture*, which at first sight is far from self-evident. In CIA manuals, torture ('interrogation under duress') is described as a technique to put the prisoner in a state of fundamental disorientation. The aim is to eliminate the possibility of resistance, something that is achieved by creating a break between the prisoner and his/her ability to understand their surrounding world. This has typically been done by blindfolding, placing a bag over the head, or by exposing the prisoner to extreme light or music, physical violence, or electroshock. Jeff Halper's study of Israeli techniques to keep the Palestinians under control speaks about how a population that can potentially resist turns into a 'malleable mass, ... a receding field in which domination becomes possible'[36]—albeit evidently as the Palestinians have proved, not sustainably.

Finally, Naomi Klein cites a Chilean psychiatrist's statement that torture under the Pinochet regime reduced adult victims to a childlike state. People became 'confused and desperate, malleable and willing to follow instructions ... they became more dependent and anxious.'[37] According to one of the CIA manuals cited, there is a moment (and it need only be very brief) when all mental activity is turned off; Klein compares this with a psychological shock or paralysis. This in turn works as a traumatic or sub-traumatic experience that blows up, as it were, the world with which the subject was familiar, including the image of himself in that world. 'Experienced interrogators recognize this effect when it occurs and know that this is the time when the source is *much more open to suggestion, will obey much more easily than before experiencing the shock.* '[38]

Now torture does not necessarily have to have the mechanical form that predominated well into the twentieth century. As early as the 1950s, the CIA in the MKULTRA project was looking for new methods of enforcing obedience, including the use of psycho-pharmaceuticals such as LSD. Under the pretext that American prisoners of war had been exposed to 'brainwashing' in the Korean War, a search was undertaken to discover psychological torture methods that centered on disorientation. MKULTRA's director, the chemical engineer Dr Sydney Gottlieb, joined the CIA in 1951; he used the wartime experience of Japanese and Nazi camp doctors and in a number of cases even brought them to the U.S.[39] I will come back to this in detail in Chapter 5.

When MKULTRA came into the public eye in the 1980s (according to Max Parry, a whistleblower had already died under suspicious circumstances in 1953), it became clear that the real goal was to refine the technique of torture. Naomi Klein quotes Alfred McCoy, the author of *A Question of Terror: CIA Interrogation from the Cold War to the War on Terror,* who writes that the MKULTRA project discovered that people could be subjected to a shock by first not letting their senses receive any stimuli at all ('sensory deprivation') and then suddenly subject them to an overdose.[40] Meanwhile, the revelations about the treatment of 'terror suspects' from Afghanistan at the U.S. base at Guantánamo Bay in Cuba and the interrogation of Iraqi prisoners at Abu Ghraib prison have made it clear that in such a treatment, physical violence and humiliation are in fact hard to separate.

The psychological shock of the proclamation of a pandemic, like the purpose behind torture, is intended to induce acceptance of a 'new normal' and to turn off critical judgment. This state of mind is achieved by withholding information about what is really going on, through the extremely one-sided information by politicians and mainstream media. Divergent views by often highly qualified experts are not mentioned or are dismissed as 'conspiracy theories.' This can be compared to the sensory deprivation in psychological torture.

On the other hand, the initially daily reporting on 'deaths' (instead of number of deceased), 'anonymous mass graves,' etc., and without ever putting the figures in perspective, evoke an intense, existential fear that is again comparable to the *overdose* of stimuli. Making face masks compulsory, social distancing and similar, medically senseless or even counterproductive measures evoke an absurd, unreal atmosphere that deeply affects people's state of mind. Research in The Netherlands revealed that during the first, still fairly restricted lockdown, in one in three people, 'mental health deteriorated during corona time due to an increase in anxiety, depression and sleep disorders.' One in ten thought about death more often than before.[41]

My contention is that the introduction of the 'new normal' in the Covid-19 crisis, although at first glance of a different order, relies essentially on the same principles and outcomes as the described techniques to induce disorientation and loss of cognitive function. The revelations about British and German plans make it clear that this was also the intent. We are dealing with a biopolitical seizure of power, initiated at the level of global governance and reaching deep into the sovereignty of the individual, a seizure that involves a whole range of forms of violence. Not least the police brutality unleashed on all continents against protesters who, for all their political differences, have one concern in common—the loss of their freedom and the desire to regain it.

It must therefore be concluded that, in light of the real magnitude of the 'pandemic,' the imposition of the state of exception in practically the entire world is primarily a political step, which, as we will see, has been long prepared and coordinated in a range of transnational think tanks, consultative structures, and formal organizations such as the WHO and the World Bank. Acting on their advice and express instructions, governments have applied a stranglehold to their populations which they cannot or dare not relax and must maintain with all means available, because ultimately what is at stake is the survival of the existing social order.

In this regard, we see a repetition of moves: many of the phenomena that accompanied the 9/11 attacks in New York and Washington and the subsequent War on Terror, with its invasions and regime changes in a range of countries in the Middle East and North Africa, are recurring in the Covid crisis. Obvious foreknowledge, a propaganda offensive, the suppression of dissent and deviant information, 'revenue models' to profit from the measures, increasing tension in society, everything has been seen before. The same with the anti-Putin campaign after the re-election of the Russian president in 2012, followed by the Kiev coup in 2014, etc.

However, the Covid 'pandemic' has left all previous fear campaigns far behind. Large parts of the world's population have been put in a permanent state of anxiety that can be stirred into panic at any moment if the rulers deem it desirable. Entire societies are being demolished. Draconic legislation suppressing elementary freedoms, the suspension of constitutions, the hostage-taking of entire populations; it all reveals something that must go to the heart of our social order for so much misery to be created in the name of a passing virus. However, the program that is being implemented (the *Great Reset* with *Build Back Better,* the *Screen New Deal,* or however else it is being labeled), has nothing to do with health. As I will show in the remainder of this book, it has everything to do with preserving the power of the oligarchy, the transnational ruling class threatened by a restless population demanding rectification of the absurd inequalities produced by a capitalist system that has run aground economically, socially, and ecologically.

Endnotes

1. F. William Engdahl, 'Can We Trust the WHO?' *Global Research,* 25 April 2020 (online); Michel.Chossudovsky, 'The Covid-19 Roadmap: Towards Global Economic Chaos and Societal Destruction (Annex: The Reverse Transcription-Polymerase Chain Reaction Test (RT-PCR),' *Global Research,* 13 October 2020 (online).

2. Patrick Henningsen, 'COVID-19: Trigger for a New World Order. Economic Stagnation and Social Destruction,' *Global Research*, 24 August 2020 (online).

3. President Lukashenko in a YouTube video by *Vagabond*, 11 August 2020 (online).

4. Flo Osrainik, *Das Corona-Dossier. Unter falscher Flagge gegen Freiheit, Menschenrechte und Demokratie* [foreword, Ullrich Mies] (Neuenkirchen: Rubikon, 2021), pp. 148–49. Magufuli's vice-president is a World Economic Forum insider; besides the 'pandemic,' the exploitation of Tanzania's nickel and other mineral deposites were at play as well. Jeremy Loffredo and Whitney Webb, 'Tanzania's Late President Magufuli: "Science Denier" or Threat to Empire?' *Global Research*, 2 April 2021 [orig. *Unlimited Hangout*] (online).

5. Figures by the respective health institutes, RIVM (The Netherlands) and FOHM (Sweden), in Ad Broere, 'Lockdown leidt niet tot lagere sterfte,' *De Andere Krant*, 3 (4), 2020, p. 5; David Klooz, *The Covid-19 Conundrum*, n.p. (Apple Books, 2020), p. 5.

6. *Bloomberg,* '99% of those who died from virus had other illnesses, Italy says,' 18 March 2020 (online).

7. Klaus Schwab and Thierry Malleret, *Covid-19: The Great Reset* (Geneva: World Economic Forum, 2020), p. 247; Kit Knightly, 'WHO (Accidentally) Confirms Covid is No More Dangerous than Flu,' *OffGuardian,* 8 October 2020 (online). Covid figures reported on *Worldometer*, 'current corona figures' (online).

8. Stadler cited in Osrainik, *Das Corona-Dossier*, pp. 156–58; *Yahoo News*, 'The PM who regrets taking tough coronavirus lockdown measures,' 3 June 2020 (online). The statement by the Milan hospital director is in the same report.

9. Chossudovsky, 'The Covid-19 Roadmap'; *Associated Press*, 'German researchers develop 1st test for new virus from China,' 16 January 2020 (online); *Wikipedia*, 'Kary Banks Mullis.'

10. Barry Atkinson and Eskild Petersen, 'SARS-CoV-2 shedding and infectivity,' *The Lancet*, 395 (10233), 15 April 2020 (online).

11. Walter Van Rossum, *Meine Pandemie mit Professor Drosten. Vom Tod der Aufklärung unter Laborbedingungen* (Neuenkirchen: Rubikon, 2021), pp. 65–69.

12. *Buitenparlementaire onderzoekscommissie,* 'Bevindingen die vernieti-gend zijn voor de corona test—moleculair bioloog Peter Borger.' *Potkaars pod-cast*, 22 November 2020 (online); Van Rossum, *Meine Pandemie mit Professor Drosten*, p. 47. The original article was published in *Eurosurveillance,* vol. 25, no. 3, on 23 January 2020. See also the *Eurosurveillance* website.

13. Osrainik, *Das Corona-Dossier,* pp. 134–35.

14. *WHO*, 'WHO Information Notice for IVD Users 2020/5; Nucleic acid testing (NAT) technologies that use polymerase chain reaction (PCR) for detection of SARS-CoV-2,' 20 January 2021 (online).

15. George Orwell, *1984: A Novel* (Harmondsworth: Penguin, 1954 [1949]), pp. 160–61, emphasis added.

16. Beverly J. Silver, 'Labour, war and world politics: Contemporary dynamics in world-historical perspective,' in Kees van der Pijl, ed., *Handbook of the International Political Economy of Production* (Cheltenham: Edward Elgar, 2015), pp. 7–9.

17. Greg Jaffe, 'A decade after the 9/11 attacks, Americans live in an era of endless war,' *The Washington Post*, 4 September 2011.

18. John Locke, *Two Treatises of Government* [intro. P. Laslett] (New York: Mentor, 1965 [1690]), pp. 421, 425.

19. Domenico Losurdo, *Contre-histoire du libéralisme* [trans. B. Chamayou] (Paris: La Découverte, 2013 [2006]). The treatment of *Wikileaks* publisher Julian Assange in this respect is an example; it is no different than the treatment of Antonio Gramsci by the Italian fascist regime in the 1930s.

20. Kevin Phillips, *The Cousins' Wars: Religion, Politics, and the Triumph of Anglo-America* (New York: Basic Books, 1999); Radhika Desai, *Geopolitical Economy: After U.S. Hegemony, Globalization and Empire* (London: Pluto Press, 2013), pp. 31–33.

21. Michel Foucault, *Sécurité, territoire, population: Cours au Collège de France (1977–1978)* [ed. M. Senellart] (Paris: Gallimard-Seuil, 2004), pp. 267–71; see also his *Surveiller et punir: Naissance de la prison* (Paris: Gallimard, 1975).

22. Carl Schmitt, *Political Theology: Four Chapters on the Concept of Sovereignty*, 2nd ed. [trans, intro G. Schwab, foreword T.B. Strong] (Chicago: University of Chicago Press, 2005 [1934, 1922]), p. 5.

23. Schmitt, *Der Hüter der Verfassung,* 4th ed. (Berlin: Duncker & Humblot, 1996 [1931]), p. 159, emphhasis added.

24. Schmitt, 'Zum 30. Juni 1934' [1940] in Léon Poliakov and Joseph Wulf, eds., *Das Dritte Reich und seine Diener* (Wiesbaden: Fourier, 1989 [1959]), p. 329.

25. Robert W. Cox, *Production, Power, and World Order. Social Forces in the Making of History* (New York: Columbia University Press, 1987); Antonio Gramsci, *Selections from the Prison Notebooks* [trans., eds. Q. Hoare and G.N. Smith] (New York: International Publishers, 1971 [written 1929–35]), p. 182 note.

26. Foucault, cited in Stefan Elbe, *Virus Alert: Security, Governmentality, and the AIDS Pandemic* (New York: Columbia University Press, 2009), pp. 59 and 79, respectively; emphasis added.

27. *Scientific Pandemic Influenza group on Behaviour and Communications,* 'The role of behavioural science in the coronavirus outbreak' (pdf), 14 March 2020 (London: SPI-B); Mike Robinson, 'COVID Coercion UK: SAGE Documents Reveal Psychological Operations Against Public,' *21 Wire*, 20 May 2020 [orig. *UK Column*, 14 May 2020] (online); *Behavioural Insights Team*, n.d. 'We're 10! Explore a decade of BIT' (online). An overview of the UK campaign can be found in Laura Dodsworth, *A State of Fear: How the UK government weaponised fear during the Covid-19 pandemic* (London: Pinter & Martin, 2021).

28. Annette Dowideit and Alexander Nabert, 'Innenministerium spannte Wissenschaftler für Rechtfertigung von Corona-Maßnahmen ein,' *Welt am Sonntag* (7 February 2021), with supplement titled '*Maximale Kollaboration*' (online).

29. Cited in ibid.

30. Angelo Giorgiani, testimony for *Stiftung Corona-Ausschuss* (Dr. Rainer Fuellmich), Session 39, February 2021, *Oval-Media* (online).

31. Philip Willan, *Puppet Masters: The Political Use of Terrorism in Italy* (London: Constable, 1991).

32. Giorgiani, in *Stiftung Corona-Aussschuss*.

33. Giuseppe Tomasi di Lampedusa, *The Leopard* [trans. A. Colquhoun, intro. David Gilmour] (London: Campbell/Everyman's Library, 1991 [1958]), p. 22. On passive revolution see Gramsci, *Selections from the Prison Notebooks*, pp. 105–20.

34. Merkel cited in Osrainik, *Das Corona-Dossier*, p. 274.

35. Naomi Klein, *The Shock Doctrine: The Rise of Disaster Capitalism* (Harmondsworth: Penguin, 2007).

36. Jeff Halper, *War Against the People: Israel, the Palestinians and Global Pacification* (London: Pluto Press, 2015), p. 180.

37. Klein, *The Shock Doctrine*, p. 111.

38. Ibid., p. 16, emphasis added.

39. Max Parry, 'Is the Global Pandemic a Product of the Elite's Malthusian Agenda and U.S. Biowarfare?' *Unz Review,* 16 March 2020 (online).

40. Klein, *Shock Doctrine*, p. 41.

41. Clara van den Berg, 'Ziek van angst,' *De Andere Krant – Covid-1984,* 3 (4), 3 February 2021, p. 5.

2.

CAN THE WORLD POPULATION STILL BE KEPT UNDER CONTROL?

O f all the factors at work in the Covid crisis, the threat of an uncontrollable world population is the most fundamental. Humanity is still exponentially growing in absolute numbers; between 1970 and our decade it doubled to nearly 8 billion. Under capitalist conditions and with digitization and automation advancing fast, there will no longer be any basis for a decent existence for many of them. However, much more important than the number itself is the fact that since 2008, humanity has also become restless, on a scale unlike anything seen before. Strikes, riots, and anti-government demonstrations have broken existing records in every category during this period.[1]

The restoration of order is therefore the fundamental explanation for the imposition, on a world scale, of states of emergency, for which all medical grounds had vanished by the summer of 2020, when the SARS-2 virus had passed. Accordingly, perhaps worn out after more than a year of whipped-up Covid panic, it appears that the public's sense of urgency is subject to serious erosion. In early 2021 Edelman, the PR company that was closely involved in the preparations for the 'pandemic,' had to conclude that the intention to use the anxiety psychosis to apply experimental gene therapy to the entire population ('vaccination') essentially failed.[2]

Meanwhile there is no letdown in the severity of 'the measures,' although cracks are opening up notably in the United States, with states like Florida and Texas and some twenty others at the time of this writing having terminated all Covid measures, and even outlawed the use of so-called vaccine passports that would entitle their holders to certain privileges in terms of basic freedoms.

In this chapter, we consider Zbigniew Brzezinski's prediction that a new '1848' (the revolutionary year in Europe, in which the *Communist Manifesto* also appeared), another widespread uprising of the peoples, was coming. Of course, democracy as we know it has *always been* under surveillance—not only by imposing a state of emergency as we saw earlier, but by permanent, albeit long-disguised surveillance which, with the pandemic, has become open. The ruling class has unintentionally created a revolutionary crisis by trying to freeze the economy and bring social life to a standstill. As a consequence of major shifts in the class structure, the traditional left associated with the labor movement has merged into the 'broad center' that implements the politics of the internationalized state (including the Covid state of emergency). As a result, it is mainly nationalist populism that has taken over the role of resistance, leading to an expansion of its influence so far, and as we will see in Chapter 3, in the United States and elsewhere, this has led to breakthroughs to power which in turn mobilized the established ruling class against outsiders, possibly even determining the timing of the emergency. The threat of further major uprisings (India, Chile, and France as examples) has engendered a militarized response to the 'pandemic.' The information warfare model developed in counterinsurgency operations in Southeast Asia and Central America is now being applied to the home front in the West.

DEMOCRACY UNDER SURVEILLANCE: OPERATION GLADIO AND THE CONTINUITY OF GOVERNMENT

D emocracy has been largely undone during the Covid crisis and a 'return' to civil life ante is not to be expected under current conditions. In any event, democracy was not a natural development, but an enforced *concession,* the extent of which was determined by the power relations between classes. In the English-speaking West, concessions were possible due to the relative wealth created by early industrialization and the super-profits made by exploiting the imperialist periphery, the colonies and semi-colonies. In Central and Eastern Europe after the First World War, on the other hand, even the state of emergency was no longer sufficient to control the strong workers' movement, with the Soviet Union in the background. That is why the most endangered fractions of the ruling class (the large landowners, the army, and obsolete heavy industry) resorted to a fascist regime in Italy and Germany to compel the population into submission again, eventually by going to war, with the paradoxical, albeit short-lived support of the population. This last aspect is again at play in the current state of emergency.

In the United States, the mass production of automobiles (including steel, rubber, glass, and other inputs) lifted labor productivity to a new level in the 1920s. When the previous industrial system, carried over from the 19th century, and overseas financial operations slipped into a crisis and the workers resorted to mass strikes and factory occupations, President F.D. Roosevelt, re-elected in 1936, had to make far-reaching concessions to avoid revolution, leading to the New Deal. The U.S. could afford them because the Ford system of production (enabling both high productivity and purchasing power sufficient to create a mass market) offered unprecedented growth opportunities at home. However, the captains of industry old and new did not want to cooperate with Roosevelt's plans until after economic policy had been adjusted to ensure that the labor market could no longer become too tight. Hence after the war emergency, the unions were brought under control by a communist witch hunt. Formal democracy too had to be carefully delineated. Making a choice from the candidates offered once every four years was good enough, according to Joseph Schumpeter in 1942.[3] While the United States had prospered, Europe had been weakened by the war. The socialist and communist movements were deeply rooted there, and in the wake of its defeat of Nazism, the influence of the Soviet Union loomed large. A Western European network of underground military units with secret weapons storage dumps was therefore installed for emergency use. After the communist takeover in Czechoslovakia in early 1948, at the initiative of the British Labour government, secret negotiations began with the United States and Canada about an Atlantic security structure. When NATO was formally established in April 1949, it assumed command of the underground network too. At the end of the Cold War, the underground network was exposed and became known under the name of its Italian branch, Gladio. Over the intervening decades the NATO underground at critical moments in different countries had influenced politics and public opinion by false flag bomb attacks, assassinations and abductions.[4]

Nevertheless, on both sides of the Atlantic, elections in the three decades after the war saw the highest participation rates thanks to the turnout of the working people. These were the 'Golden Years' of collective wage bargaining, price support for agriculture, and year in, year out improvement of overall conditions in the workplace and in society at large. All this enabled the Cold War and neocolonial adventures in Southeast Asia, the Middle East, and elsewhere, with mainstream Social Democracy or its equivalents outside Europe acting as brokers.[5]

In the late sixties and early seventies this social contract began to unravel because of labor shortages that further empowered workers. Also, the post-war generation was rebelling over U.S. warfare in Southeast Asia, then still extensively reported on TV, with further turmoil arising due to the denial of civil rights to blacks in the American south, and the emergence of a new youth culture. The ruling class now began looking for ways to roll back

democracy, measures that obviously could not permanently take
the form of murdering politicians deviating from the Cold War
course, such as had occurred in the United States with the assassinations of the Kennedy brothers (the president in 1963 and his
brother Robert in '68) and the black civil rights leader, Martin
Luther King, Jr., also in 1968.[6] In Italy, Aldo Moro was murdered
in 1978 after having been warned several times not to allow the
left a role in government; in West Germany, Chancellor Willy
Brandt, the architect of East-West rapprochement, was sidetracked
by a spy scandal, to name only the best known cases. Operation
Gladio, applied in European countries whose populations were at
risk of harboring communist inclinations through the Strategy of
Tension, intended to frighten a population with the paradoxical effect of drawing them closer to the government. Terror attacks were
among the tactics used to evoke fear, and the NATO underground
was a major contributor to the violence.[7] This was one of the reasons why the independent-minded French President, De Gaulle,
withdrew his country from the military organization of NATO,
the headquarters of which was subsequently forced to move from
Paris to Brussels in 1966.

The seriousness of the democratic challenge called for a
more thoughtful and sustainable approach. In 1975, a report
on the topic was published by the Trilateral Commission (TC),
a new private consultative body that had branched off from the
Bilderberg Conferences a few years earlier. In such networks,
prominent figures from North America and Western Europe, in
this case expanded to include political leaders and businessmen
from Japan (today from all over Asia), come together to consider
possible answers to problems demanding a joint response. Like all
Trilateral Commission proposals, this report was intended to give
the proposed 'new normal' a technocratic, even superficially progressive tone in order to cover its plainly reactionary agenda. The
key contributor, Samuel P. Huntington, and his co-authors (for the
other two regions represented in the Commission) noted that 'in
recent years, the operations of the democratic process ... appear to
have generated a breakdown of *traditional ways of social control*,

a de-legitimization of political and other forms of authority, and an overload of demands on government, exceeding its capacity to respond.'[8] In other words, democracy was getting out of hand.

The report still left a range of issues unaddressed but elements of the neoliberal new normal were already detectable in it. Thus, one sure way of reducing the demand load on government was to bracket off the structure of the economy from what could be legitimately decided by elections and parliamentary procedure. Others had already explored issues such as the need to abolish generous social services including student grants, in order to limit the opportunities for people to become politically active at no direct cost to themselves.[9]

In 1978, Huntington became security coordinator in the administration of President Jimmy Carter, who had assumed office in January of the previous year. Together with Brzezinski, who had been involved in launching the TC and who was now Carter's National Security Adviser, he went to work to revamp the crisis planning system, in particular by the creation of the Federal Emergency Management Agency (FEMA). FEMA was tasked with setting up the infrastructure for emergencies, including natural disasters.[10] This significantly expanded the existing program for instituting the state of emergency in the United States, the system of Continuity of Government, COG for short.

COG was originally introduced under President Eisenhower in the 1950s and had been intended to ensure the continuance of the government after a nuclear attack. After the large-scale riots that followed the murders of Robert Kennedy and Martin Luther King, Jr. in 1968, it became part of the infrastructure for the restoration of order after major disturbances, at the express demand of the U.S. military. As we will see later, it would also be activated in the Covid crisis. Paradoxically, Richard Nixon, elected president as a law-and-order candidate confronting the forces mobilized by the broad anti-war and equal rights coalition, also came into conflict with the CIA and other elements of the national security state. The U.S. military, in particular, targeted him and his advisor and later Secretary of State, Henry Kissinger, over détente with the Soviet Union and the rapprochement with China that had been kept secret from the Pentagon. Despite his resounding reelection in 1972, an indication of the appeal of a law-and-order policy for an electorate facing social unrest and significant cultural change, Nixon was shortly forced to resign due to the Watergate scandal.[11]

Proponents of a tougher stance, both domestically and in foreign policy, grew stronger under Nixon's successor, Gerald Ford. Donald Rumsfeld, who had been ambassador to NATO under Nixon, became defense minister; his assistant, Congressman Dick Cheney, was made White House Chief of Staff. At the same time (1975–76) CIA estimates of Soviet weapons outlays were revised upwards and the (fabricated) figures used for an alarmist media campaign.[12] After the Carter interregnum, the neoconservatives (Neocons) broke through under the presidency of Ronald Reagan, who then unleashed an unprecedented peacetime arms race as well as attacks on the trade unions. Rumsfeld and Cheney (who otherwise held no office in the Reagan administration) were now tasked with further expanding the COG system to allow the military to take over in a state of emergency.[13]

Huntington, meanwhile, in a 1981 book, *American Politics: The Promise of Disharmony,* argued that the United States was ill-prepared for social unrest. Europe, he maintained, had the advantage of having a much longer experience in dealing with the workers' movement and socialist forces, and this precisely was lacking in the U.S. There was no clear understanding of the role of the state either; for most of its history, the country had never been subject to properly centralized authority, nor did it possess the experienced bureaucratic machinery of the main European states. Historically, the United States political system was therefore weak and vulnerable to insurgent movements.[14]

The overhaul of the COG infrastructure by Rumsfeld and Cheney was to make up for this shortfall. Extensive facilities for mass surveillance and detention of dissidents in camps were planned, as well as the appointment of military commanders to exercise authority under exceptional circumstances. Besides directing the Contra operation in Nicaragua, Vice-President George

H.W. Bush and his White House basement confidant, Colonel Oliver North, worked with FEMA to perfect the COG system into a shadow government that could take over from the formal, statutory agencies. With a system of command centers spread across the United States, some of them mobile, and with a communications and logistics headquarters in Arizona, the country could be run under a state of emergency, if need be. The command centers were connected to four Boeing E-4 command posts, housing so-called Doomsday aircraft, stationed at Offut Air Force base in Nebraska.[15]

This parallel state sprang into action on 9/11, with Cheney now as vice-president in charge, and Rumsfeld again as defense secretary back in the Pentagon, surrounded by Neocons like his deputy, Paul Wolfowitz, and others. In the ensuing War on Terror, as Adam Curtis shows in his 2004 BBC documentary, *The Power of Nightmares*, the promises of the American dream soon evaporated, to be replaced by the politics of fear, a mode of government based on scaring the public into submission. With the Patriot Act, democracy was rolled back by several degrees. Orwell's assessment of permanent war as the means to safeguard the existing social order had truly come into its own. Only with the *Wikileaks* revelations about U.S. war crimes and Edward Snowden's about the wiretapping of the entire world by the Five Eyes (the signals intelligence services of the U.S., Great Britain, Canada, Australia and New Zealand) was the full extent of this system made visible for all to see.[16]

With the advent of the Covid crisis, the parallel or 'deep state' system described above is going through another major overhaul, this time as part of a global information war against the world population. The establishment of an authoritarian surveillance society presupposes that the liberal division of powers between the legislative, executive, and juridical arms of the state, be replaced by an integral state power that abolishes existing forms of popular sovereignty and subjects the population to permanent control.

POPULATION SURPLUS IN DIGITAL CAPITALISM

Since the 1980s, the attack on the social state in the name of neoliberalism in practically every country of the world has removed the buffer that shielded populations from their unrestrained exploitation by capital.[17] The biggest shock in this transition was the opening of China and the collapse of the Soviet bloc and the USSR, which increased the global labor supply from 1.5 to more than three billion people in two decades.[18]

Because of the simultaneous breakdown of social protection in the richer parts of the world (albeit less shock-like), social insecurity has increased everywhere. This not only produces 'rational poverty,' i.e. a shortfall in purchasing power, but specific psychological disorders as well.[19] If this is already happening in the still relatively prosperous parts of the world, what about the less fortunate regions? In those areas, indeed those further removed from the capitalist heartland, the exhaustion of the social and natural foundations of organized existence, as well as armed conflicts that flow from it over raw materials and other sources of wealth, lead to extremes of poverty and destitution.

At the end of the twentieth century, the International Labor Organization (ILO) estimated that about one-third of the global labor force was unemployed or only partly employed. Around the time of the 2008 financial collapse, it was estimated that of non-agricultural labor, 82 percent was 'informal' in South Asia, two-thirds in Sub-Saharan Africa and in East and Southeast Asia, and 51 percent in Latin America. The global supply chains along which cheap industrial goods are transferred to developed economies, originate in those unregulated labor markets.[20] In 2011, there were 1.53 billion workers worldwide in unstable employment relationships, half of the total. In 2019, *the majority* of the 3.5 billion workers were in this condition. A UN report predicted that the emergence of online labor exchanges would further deepen the downward spiral of employment conditions.[21]

This trend is also increasingly evident in the advanced industrial countries. Since 2008 the new wave of digitization and the emergence of IT platforms have triggered a very rapid trans-nationalization of services. In 2017 they accounted for about 70 percent of the total gross world product. As a result, even highly skilled knowledge work is being replaced by computers, and a fusion of new technologies is occurring as the boundaries between the physical, digital and biological worlds are fading.[22] In a WEF White Paper, the aforementioned World Economic Forum has drawn the contours of the job market expected in 2030. It estimates that 83 percent of the workforce will be working from home and 40 percent of all training and education will also be digitized, so that it can be organized at a distance. This largely suspends the social dimension of those activities (work and education), resulting in the effective isolation of the vast majority of people, confined in their workstations at home. Between 13 and 28 percent of the world's population will become temporarily or permanently redundant, say, 1 to 2 billion people who no longer have a role to play in the process of social and economic reproduction.[23]

At the same time, wealth accumulates at the top. 'Capitalism' today is in actuality an oligarchy of multi-billionaires holding power, assisted by a cadre of relatively highly trained and paid specialists. In his bestseller, *Capital in the 21st Century*, Thomas Piketty speaks of an oligarchic trend, 'a process whereby *rich countries become the property of their own billionaires.*'[24] This trend is also evident in China and Russia, although the state class, which bases its power not on property but on its grip on the state apparatus, still retains an influence at least equivalent to that of the oligarchs.

A NEW '1848'?

The people made redundant by digital capitalism will not remain passive, waiting for what will happen next. They have already begun to respond actively, through all kinds of resistance and through migration from the poor parts of the world to the wealthier ones, from the countryside to the cities. Since the 1980s,

a majority of the world's population has been living in urban set-
tings and unrest has become endemic there, too. One consequence
is that conflicts that used to be fought in the countryside or in
the jungle, such as in Indonesia, Vietnam or Central America,
have now been urbanized. The dividing lines between strikes,
demonstrations and street riots with arson and looting have also
become blurred. A 1997 report by the RAND Corporation speaks
of an 'urbanization of uprisings.' It does not matter whether we
are dealing with politically conscious opponents, with 'terrorists,'
with ordinary street crime, or with aimless unrest. Jeff Halper
quotes an Israeli population control specialist who dismisses the
emphasis on terrorism, contending the issue is rather how to deal
with a 'restless underclass.'[25] 'The real War on Terror,' writes
Mike Davis, is 'the low-intensity world war of unlimited duration
against criminalized sections of the urban poor.'[26]

Huntington, in his *Clash of Civilizations,* which originally
appeared as an article in *Foreign Affairs* in 1993, pasted the label
'Muslim' on the global labor surplus and saw the Middle East
as the main hotspot. Western civilization, he argued, was facing
the challenge of the combined threat of Islam and 'Confucianism'
from Asia. He also viewed Islam as a source of inspiration for
violence, read: terrorism. The young, unemployed population in
Muslim countries, who can only hope to find work in the cities,
including those in the West, have brought this violent doctrine
into the heart of Christian civilization—at least according to
Huntington's thesis.[27] In turn, as a result of the general feeling
of insecurity due to the many newcomers in the expanding cities,
there has also occurred a militarization of the police force and
the level of violence has generally increased. In the process, law
enforcement has adopted certain traits of gang life, with police
actions becoming more brutal as they assimilate the practices
of street criminals, no longer covered by the rule of law but by
sheer force. Israeli expertise in handling Palestinians serves as the
training manual here.[28] In the Covid crisis, many people now have
first-hand experience of this sort of behavior.

Huntington's views on a threat from the Islamic world dovetailed with efforts to have the United States wage Israel's wars with its neighbors billed as a War on Terror. Benjamin Netanyahu, the future Prime Minister, at one of the preparatory conferences where this theme was developed (this one in Washington in 1984) spoke of a 'population united by fear.' This is now intended to extend beyond the period in which an incident occurs; a permanent state of emergency is being established in which the difference between normal periods of orderly coexistence and situations of actual, often extreme violence is eliminated.[29]

In November 1999, however, a specifically anti-capitalist movement also emerged, when thousands of protesters from around the world gathered in Seattle on the American west coast to manifest their resistance against further liberalization and internationalization of the world economy.[30] Certainly the 'anti-globalization movement' that burst onto the scene in Seattle would be sidetracked by the explosion of violence after 9/11, but the threat of a global uprising, and the possibility that it might explicitly turn against the capitalist order, did not dissipate. Ultimately, it does not matter to the Western ruling class whether resistance stems from organized labor unrest or from religious or ethnically inspired discontent, and the reliance on army and police violence as a result has slowly but surely become routine.[31]

The financial crisis only accelerated the social struggle on a global scale. In 2008, serious disturbances broke out in more than twenty countries because people could no longer pay for their daily purchases. World food prices had first gone through the ceiling as speculative practices got out of hand; when they collapsed, prices in many poor countries remained high as supplies dried up. Leasing agricultural land in poor countries to ensure food security in rich countries has only exacerbated local problems.[32]

As mentioned, in the years that followed, *all indicators of social unrest around the world showed an upward trend.* A comparison of the figures in the Cross National Time Series (CNTS) shows that every record for social unrest has been broken subsequent to 2008. There was a sharp increase in the number of strikes after

2011, when the number tripled in one year after years of decline; in 2015 the previous record (1988) was broken. Anti-government demonstrations also rapidly increased in number after 2010, and there was a no less spectacular increase in the number of riots (a six-fold increase after 2011), breaking the record of 1968–69 in 2013. Trust in government, and even more in information, declined in all countries.[33]

The uprisings in Tunisia, Egypt and elsewhere in North Africa and the Middle East (the 'Arab Spring') prompted a warning by Brzezinski that the global population surplus in combination with the Information Revolution heralded *a new 1848*. It might well be, he argued in his last book, *Strategic Vision*, that the millions of young people in the world today are the equivalent of Marx's 'proletariat'; while their prospects for a dignified existence are minimal, they are abundantly informed through the Internet about the political and social reality that causes their plight.[34] Others too recognized the similarity with 1848, but in a 2013 report the International Labor Organization (ILO) concluded that the most acute danger of such an explosion lay not in the Middle East but *in the EU*, and there, in the southern half more than in the north.[35] France in particular appeared to be heading for a revolutionary crisis—until the 'pandemic' was declared. Before going into that, we should consider the changes in the class structure of Western society more generally.

MIGRATION AND THE METROPOLITAN UNIVERSE

The redundancy of more than a billion people under conditions of impoverishment, conflict, and the collapse of natural substratum due to over-exploitation and pollution, is of course much more explosive in the formerly termed Third World than in the West, which still has some reserve in this regard. Just as Africa had once been declared 'under-polluted' by U.S. Treasury Secretary, Larry Summers, and therefore a good dumping ground for chemical waste, continental Europe, according to another leading business strategist, Irishman Peter Sutherland, is 'under-populat-ed' by non-Europeans.

Sutherland occupied several key directorates such as chair-man of Goldman Sachs International, BP, and other companies, besides serving as the UN's High Representative for Migration towards the end of his life. In 2012, in a series of lectures and interviews, he held up the English-speaking West as an example to Europe in matters of migration. The EU should open its bor-ders completely because the U.S., Canada, Australia, and New Zealand, according to Sutherland, also owed their prosperity to this. As structurally immigration countries they much more eas-ily assimilate additional newcomers, irrespective of their diverse backgrounds. In contrast, 'we [as Europeans] still have a sense of our homogeneity and difference from others,' and Sutherland's urgent advice was that the European Union should do its utmost 'to undermine that.'[36]

At the UN and as head of the Global Forum on Migration and Development until his death, Sutherland worked consistent-ly to make migration a generally accepted and normal form of distribution of the population surplus across the continents. In December 2018, his efforts were rewarded when 164 countries in Marrakesh signed the Global Compact for Safe, Orderly and Regular Migration. On that occasion, the UN Secretary-General, Antonio Guterres, called migration inevitable and necessary; the agreement declares migration a fundamental right that should be honored with full access to public services in the countries of arrival, regardless of the status of the migrant. Incidentally, under this (non-binding) agreement, the signatures of two of the immigration countries recommended by Sutherland, the U.S. and Australia, were missing.[37]

An important explanation for the Anglophone settler societ-ies' ability to absorb multiple waves of immigration, resides in the fact that the indigenous populations of North America, Canada, Australia, and New Zealand, viz., Amerindians, 'First Nations' in Canada, Aborigines, and Maoris, respectively, were largely displaced, decimated and even exterminated. Certainly, there are evident hierarchies between the original Anglo settlers and later arrivals, but in the end none of the immigrant categories is in a position to claim they are the 'original' inhabitants, since only remnants of these survive on the fringes of society, in reservations, etc. In addition, of course, in the United States, the descendants of the slave population brought from Africa, whilst formally emancipated after the Civil War, under the Reconstruction after the victory of the North, were relegated to second-class citizens again by Jim Crow laws in the South. Despite attempts in the 1960s to accommodate the black civil rights movement by according them civil rights, African Americans' second-tier status has remained in many ways, cementing the unity of the white population, as well as Hispanics to some extent.

It is not easy to reconstruct the political motives behind the promotion of migration apart from mobilizing the reserve army of labor. It would seem that the one possible rationale underlying the vision of Sutherland and the UN (which, as we will see, was placed under the supervision of transnational capital through a prior *Global Compact* in the 1980s) resides in the attempt to dissolve the social cohesion of European societies. After all, in Europe, the 'natives' have managed to hold their own, because the immigrants have come in predominantly as a labor reserve, not as armed colonizers. But this cohesion also makes them a potential obstacle to raising the rate of exploitation on which capital accumulation relies. (This social cohesion is also one target during the current Covid crisis, pursued through the medically meaningless 'social distancing' requirement). Another possible consideration is that the social and security infrastructure of the West is better able to deal with social unrest than the emigration countries, so there is also a control aspect in migration, given that it will always be the younger, most active and enterprising (and potentially troublesome) people who migrate. Even so, the survival of the bulk of the indigenous population in Europe remains an issue of major importance. Its resistance to mass immigration so far has mainly assisted the rise of national populism.

CADRES, IMMIGRANTS AND THE DOMESTIC SURPLUS POPULATION

The result of the distribution of the global population surplus over the cities (the result of *all* migration is urbanization) is that an urban immigrant sub-proletariat has formed in the West, which has reached a certain modus vivendi with the privileged layer of the technical and managerial cadre, also concentrated in the cities. This modern service category, comprising up to 20 percent of the working population in countries such as France and the Netherlands, shares the urban space with an immigrant population in the lowest income class, which has irregular work, poor housing, etc. In eastern Europe, on the other hand, immigrants have been largely kept out, and conservative leaders have successfully built their constituencies among the 'indigenous' population, which in western Europe feels forgotten and abandoned.

In his book *La France périphérique* of 2015, Christophe Guilluy compares the demographics of the French metropolises— Paris, Lyons, etc.—with the provincial towns. He demonstrates that the latter, making up the 'periphery,' are typically inhabited by a marginalized 'native' population whose economic role has been greatly devalued.[38] In other countries, there is a comparable distinction: in the U.S., the Midwest serves as the periphery of the big cities of the east and west coasts; in Britain, northern England is the periphery of London and the south-east, and so on. Even in The Netherlands, where there is an almost integral conurbation in the west country (Holland), the eastern provinces constitute a periphery in Guilluy's sense, although the dividing lines are less sharply drawn. The periphery everywhere is dominated by former workers, office workers, shopkeepers, and farmers, who are marginalized relative to the big cities. So, who lives in these?

The oligarchy as such does not count numerically, ever less than heretofore in fact, although its influence on society is enormous, as we will see in Chapter 3. But the upper layer of qualified cadres in the urban agglomerations accounting for 20 percent of the population occupy a prominent position in key respects. They earn high salaries in the service sector, in design and consultancies of all kinds, at banks and insurance companies, investment firms—either as employees or self-employed. There also remains a technical and managerial cadre in modern industry, which is increasingly connected with all kinds of intermediary companies; these too more and more resemble the service sector. In addition, there are the lawyers, doctors, and other liberal professions. All of these are increasingly internationalized, so they are used to communicating with partners abroad, to foreign travel for business and training purposes. There are also many 'expats' working for foreign companies in the cities, adding to their cosmopolitan profile (with its specific consequences for the housing market).

Paris and the other major French cities, then, focus on design and research, the knowledge economy, governance, finance, etc. Although Paris is home to only 18 percent of the French population, it contributes 30 percent to the added value of the French economy, and with three other metropolitan agglomerations, more than half.[39] It is no different in the big cities of other Western European countries, or in North America and East Asia. In the author's city of residence, Amsterdam, the make-up of the workforce is typically skewed towards 'consultancy and research' (one-fifth of the total employed population), financial institutions, information and communication, and so on. The labor market in the metropolitan service sector is very dynamic; incomes in the knowledge economy are relatively high, and housing is a good investment. Not all of the urbanites are highly paid of course: health and welfare, the second-largest category in Amsterdam, also contains a large low-paid component. But politically and culturally, the privileged cadre dominate the big cities in the West.[40]

The mainstream in politics and the regular media, notably the news programs and the talk shows that provide the key issues of the day, reflect the life-world of the large urban agglomerations. The matters which preoccupy the qualified cadres also dominate the media and politics. Concerns about the deterioration of the environment and lifestyles that take this into account, such as veganism, renewable energy, etc., resonate in public debate. Incomes are typically individually negotiated or improved through *job-hopping*. The intellectual legacy of the workers' movement, on the other hand, from wages and strikes to the struggle against imperialism, means little here. Instead, support for 'humanitarian intervention' is easily mobilized, albeit with a high 'cookies for Sarajevo' content (i.e., symbolic support for the illegal NATO bombing of rump-Yugoslavia in 1995 and 1999 and later repeated in the seizure of power in Kiev in 2014). Moral concerns have displaced historical materialism in analyzing world affairs altogether, geopolitics is a matter of the good (us) against the bad and the ugly (them). Hence also the fashion of 'apologizing' for slavery and other historical human tragedies in which we unfortunately were *not* the good ones we claim to be today.

The themes that occupy the urban cadre broadly come under the heading of *identity politics,* diversity in particular, both sexually and ethnically. The spread of more and more taboos by altering terminology, often by subtle shifts, disseminated via universities and media, sustains the 'right tone' from which it is better not to stray.[41] For the cadre, life is all about choice. Tolerance of difference is not enough: being male or female as such (or of course, transgender) is a choice too, one that schoolchildren must already be presented with.

The celebration of diversity includes ethnic difference. Anti-racism is the starting point. Not so much out of material solidarity with anyone, but to avoid offending the immigrant and minority populations with which one has *to share the urban space*. A repetition of the riots that have bloodied America's major cities and afflicted London and Paris in the recent past must therefore be avoided at all costs. The Western metropolitan anti-racists are usually less concerned about the three and a half million Congolese who perished in areas where the coltan for the cadres' cell phones is being mined.

Continuing immigration will not be a concern for the technical and managerial cadre either, since they compete for work and living space only with their own group. Basically, the urban upper layer has no affinity whatsoever with people who don't earn their living from behind a laptop, nor does it sympathize with what they undergo during the Covid lockdowns, for example.[42] This has created a widening gap in the trust that rulers and media still enjoy among the 'better informed,' their cadre, and the mass of the population, both the urban immigrant population and the marginalized 'natives.' According to the Edelman trust barometer, in France, 65 percent of the 'better informed' cadre still trust the government and the mainstream media, against 45 percent of the mass of the population; in the United States, 62 percent of the former group, against 44 of the population at large. In Britain the percentages are 59 and 43, respectively, and in Germany 62 and 52. If Edelman is to be believed, in both Russia and China, government and mainstream media have little credibility left whatsoever.[43]

After the departure of industry to low-wage overseas locations, the immigrant sub-proletariat moved into the former working-class districts. More than half of the residents of such neighborhoods in France are now immigrants, whose incomes have fallen sharply along with the quality of housing. People with a migrant background today constitute the majority in big Western European cities.[44] For them too, there is a functioning labor market, with jobs on offer all the time: as cleaners, in the catering industry, if necessary in the criminal sphere. For those who do not want to step outside the bounds of legal activity, there is a social benefit or the local family network, because for various groups of immigrants a substitute society is available in the new place of residence, with their own qualified protectors, and via the satellite dish, mosque etc., their own subcultural enclaves.

The question now arises: what has happened to the *political legacy* of the indigenous working population that due to the loss of industrial employment and the decline of the associated middle class has been marginalized by this urban combination of qualified knowledge economy and underclass? Or the shopkeepers as a result of the advance of large retail chains?

My thesis is that the decline of the traditional left as a life-world has allowed its former political representatives (working-class parties, trade unions) to merge into what I call the *broad center* of politics. Within the broad center, in which the positions originating in the metropolitan universe dominated by the cadre are hegemonic, they constitute the center-left, often in coalitions with the center-right. Beyond the (sub-) cultural positions outlined above (diversity and identity politics), this concerns their acceptance and even active promotion of neoliberal, globalizing capitalism, including its 'police' dimension—the humanitarian intervention that used to be labeled imperialism and gunboat policy. As a result, the only political current still able to cater to the discontent of the marginalized, is nationalist populism, which we will discuss at some length in the next chapter. Hence the segment of the population that is being displaced by the social forces associated with globalization at both the upper and lower extremes, tends to be dismissed as racist and 'rightwing,' even 'extreme right,' by the broad center as soon as it raises its voice. In turn, the broad center with its huge political and cultural influence, is being labeled 'leftwing,' even 'far left' by the spokespeople of national populism, though it is nothing of the kind. Again, this confusion of labels will be elaborated in Chapter 3, where I will argue that nationalist populism is in fact supported by the ruling class too but kept at a distance by the mainstream, for reasons explained there.

Yet in several countries the protest movement has taken undeniably progressive forms. I would go so far as saying that if this had not been the case, the need to pull the Covid-emergency brake might not have arisen in the first place.

UPRISINGS IN INDIA, CHILE AND FRANCE

We have already seen that the rapidly growing world population in the years following the financial collapse of 2008, just as on the eve of the two world wars, became restless, and on an unprecedented scale at that. Concrete demands were going in all directions, if they were made at all; from wage struggles and anti-capitalist movements to separatist drives as in Scotland and Catalonia. But riots and looting without a clear political signature were also part of mounting tide of popular discontent and anger.

However, a new '1848' refers to more than that. It suggests progressive social movements with inclusive, non-sectarian demands, from incomes and housing to democracy. Such movements were in evidence in several countries.

I start with India. Here Narendra Modi and his Hindu nationalists, an effectively fascist movement, had come to power. Modi had been elected in 2014 thanks to his personal popularity and by mobilizing a sectarian Hindu undercurrent against the country's Muslims. This had a long legacy, dating back to British colonialism, which had relied on communalism all along, though under the Westernized Congress rule this had never entirely displaced the syncretistic Hindu tradition.[45] Modi, however, played all registers of anti-Muslim sentiment in the country whilst continuing to implement the neoliberal turn that caused unprecedented, literally suicidal misery at a mass level, especially among the peasants.[46] However, in the run-up to the spring 2019 elections, his shine began to fade and although he held out, this was obviously the time to play his last cards. In December of that year, Modi abolished the autonomy of Kashmir province with its large Muslim population; he also had an immigration law enacted which awarded citizenship to newcomers unless they were Muslim. Throughout India this led to massive demonstrations and disturbances in which other issues too found expression. This was politically translated when Arvind Kejriwal led his party of the poor, AAP, to a sweeping electoral victory in the capital Delhi in February 2020. Squarely against the anti-Muslim policy of Modi's BJP, the winning slogan was: the real nationalism is committing yourself to the people.[47]

The cautious conclusion may be that this revival of the social struggle was in no way destined to enter national populist waters; on the contrary, the anti-Muslim populist was already in power. Certainly the 'pandemic' came just in time for Modi. India registered 97 deaths attributed to Covid-19 per million inhabitants, totaling nearly 160 thousand people in a country where 7.2 million of its 1.4 billion inhabitants die every year.[48] The country nevertheless went into a strict lockdown on 24 March; government officials appeared on TV wearing face masks and at a generous distance from each other, etc. For the mass of the population, the theatrics were less enticing: many hundreds of thousands of domestic servants were sent away for fear of 'contamination' and had to return to their villages on foot. The country was frozen into political stalemate and in April 2021 even sounded the alarm over a new Delta wave of the 'pandemic,' which occurred on the heels of a vaccination campaign (even though the government briefly before had begun distributing cheap packages of the highly effective Ivermectin).

Let us now look at Chile. Around the turn of 2019–20, this country too was rocked by huge demonstrations. This was an uprising of great symbolic significance, because in 1973 a bloody military coup d'état had put an end to the socialist government of Salvador Allende. Thereafter, the country had the dubious distinction of becoming the first testing ground for neoliberal 'market reforms,' intended to put an end to the integrated industrial economy and labor movement forever. However, in October 2019, student riots broke out to which the government of billionaire President Piñera responded with military force. The ensuing broad protest movement proved difficult to contain. When, after months of protests, there were already 26 deaths, several of them in police custody or at the hands of the military, the government seemed ready to make concessions, but this obviously came too late. Only a new constitution would satisfy the public and a referendum was planned for 26 April 2020, in which the population would be asked what form a Constituent Assembly should take.[49] A preliminary plebiscite showed there was unprecedented support for a new constitution—but here too the Covid 'pandemic' intervened just in time. A national curfew was imposed on 22 March to contain the outbreak (with 632 reported *cases* in Chile at the time). Briefly afterwards a lockdown was imposed on Santiago and other cities. Meanwhile the country is one of the few to date where an official immunity card system has been introduced.[50] The elections for a constitutional convention were postponed to May 2021, completely overturning the balance of forces in favor of the Left.[51]

Now then let us turn to France, which is the real hotspot of popular discontent in the EU. It was also because of France that the ILO in its 2013 report saw (southern) Europe as the center of a new 1848. It explains why the EU has played such a prominent role in preparing for the 'pandemic' and the vaccination campaign, as we will see below. My thesis is that the French movement cannot be reduced to populism either; it is only because of the lack of a truly progressive alternative that support for Marine le Pen is so strong. Jean-Luc Mélenchon's, *La France insoumise*, that promised a 'France that will not be subdued,' failed to consolidate his surprising success in the 2016 presidential election. In addition, Marine Le Pen's party, unlike nationalist populism elsewhere, does not have a neoliberal social program. It has broadly adopted the economic policy of public investment and purchasing power that was also backbone of the *Programme Commun* of the united left of the 1970s.

Guilluy, the author of the aforementioned *La France périphérique*, of course was a much sought-after commentator when the weekly demonstrations of the Yellow Vests started at the end of 2018, triggered by an increase in the price of diesel. After all, it was not difficult to see that the marginalized 'indigenous' French from the provinces were demonstrating here. Their slogan was '*on est là,*' 'we are (still) here.' However, the movement was not supported by the trade unions, and even the CGT leadership held back despite pressure from members—until the nation-wide revolt over the individualization of the French pension system with its implications for more inequality merged all social protest into a single movement.[52] The urban cadre, however, held aloof. When the daily, *Le Monde,* contrary to the habits of the mainstream media, did report a demonstration of the *gilets jaunes,* reflecting on its backgrounds, a vitriolic reaction from Parisian and other metropolitan readers ensued. Even the paper's editors were taken aback by the aggressive disdain for the demonstrators. That didn't stop Guilluy from predicting that the *gilets jaunes* would be unstoppable.[53] That was well before the Covid 'pandemic,' which has since evolved into a new stage of the ongoing class struggles in France. Whilst 'the measures' are adding another million to the already 5.8 million registered jobseekers, the deployment of the army is a sign that we are dealing with a coup d'état in the classic sense of a shock-like reinforcement of state power relative to society.[54]

France has experienced intense social unrest due to the policy of de-industrialization and the sustained demolition policy applied to the social security system by successive presidents, starting with the Socialist Party's François Hollande (with Macron as adviser, then minister and now with Macron as president). Still in November 2020, in spite of Covid restrictions, a mass demonstration was held against the security law. This built on movements against the Labor Code from 2016, and then, that of the Yellow Vests, all culminating in the battle against the pension reform. The latter was assuming the characteristics of a general uprising at the turn of 2019–20. In between there were an array of local movements against factory closures and layoffs. For all their differences, these popular movements, according to Claude Serfati, are an expression of the social exhaustion affecting France and to which the answer has been the Covid state of emergency. In addition, the Macron government has introduced a registration system that records a person's political, philosophical and religious views as well as union membership.

The security law with its sweeping provisions in a way contradicts Guilluy's thesis about peripheral France, in that law enforcement is now being geared against a continuum of presumed threats, from social movements to outbursts of anger in the suburbs. Both the supporters of the Yellow Vests and the immigrant urban population are targeted. France, according to the minister of education, is facing the threat of *Islamo-gauchisme*, the left plus Islam, a combination claimed to be rampant in the universities especially. The law against 'Islamist separatism' more particularly is intended to counter what is regarded as an effort to create a 'counter-society,' rejecting the country's strict form of secularism.[55]

Hence, as in India and Chile, the target of state power is the progressive movement, and that obviously is never a matter for France alone. After all, there is much to be said for the thesis that Germany, as the most powerful country in the EU and most important foreign investor in France, is closely involved in the counterattack via the German-dominated EU. As in 1870 and 1940, it can intervene (at that time, militarily) in the knowledge that on the other side of the border a substantial part of the French ruling class is ready to welcome German support in suppressing popular revolt.[56] Here too, the 'pandemic' arrived just in time, that is, at the height of the pension reform struggle in which the Yellow Vests also joined. As early as January 2020, it was clear that Macron's 'party' would be hit hard in the spring municipal elections; as soon as 'pandemic' was declared, the president accordingly did not hesitate to postpone the second round of the election. When it was finally held, the rate of abstinence was even closer to 60 percent than in the first round, as disillusion with politics spread further.[57] This brings us to the actual use of military force.

INFORMATION WAR AS A BY-PRODUCT OF COUNTERINSURGENCY

The Covid crisis has been seized on, if it was not actually unleashed for this purpose, to restore discipline within the population through a fear-based information warfare campaign. The techniques for such a campaign had been developed in U.S. counterinsurgency operations. As with the comparison with torture, this at first glance seems a tenuous connection, but there runs a direct line between imperialist counter-guerrilla warfare in Vietnam, Central America, Afghanistan and Iraq, and the repression currently being applied through the lockdowns.

The struggle for the decolonization of Africa and Asia was about preventing independence from European mother countries from being won by groups unwilling to submit to a neocolonial relationship. After decolonization, the Americans, who oversaw this strategic goal, adopted the tactics of the British in Malacca and Kenya of sending agents to penetrate the resistance and then, on the basis of centralized information, moving to take out the leadership. After the success of the bloody military coup against Sukarno in Indonesia in 1965, involving death lists provided to military commanders and radical Muslim students, the Americans launched Operation Phoenix to track down leading cadres in Vietnam as well. They were murdered by the tens of thousands in order to eradicate the resistance at the root.[58]

The Phoenix model, with its penetration and provocation based on systematic intelligence gathering, was further elaborated in the Strategy of Tension in Italy in the 1970s and in a series of illegal drug and anti-terrorist operations in Lebanon and Central America in the 1980s. Following the Six-Day War of 1967 in which Israel conquered further large swathes of Arab land it still holds today (except Sinai), the Zionist state also developed its own tactics of using double-agents and targeted assassination of Palestinians. U.S. and Israeli agents and advisers in turn were deeply involved in the death squads in Guatemala and El Salvador, fighting guerrillas in those countries.[59]

Israel's experience with the information-liquidation model gained a great reputation in intelligence circles; when the outgoing chief of the country's Internal Security Service was asked in 2005 if he had any qualms about arbitrarily murdering opponents, he replied that foreign delegations came to Israel every week to be taught these highly effective methods, and not only from the United States. After all, Israel had elevated 'targeted prevention' to an art form; it eliminates the experienced leaders, and the novices taking their place are much easier opponents.[60] This has evolved into the 'global control matrix' referred to earlier, which has entered a new and decisive phase with the Covid crisis.

After 9/11, U.S. special forces were tasked with infiltrating terrorist networks and 'stimulating responses,' provoking them into action. In 2005, journalist Seymour Hersh revealed that they systematically worked with their own terrorist units that could trigger violence by others in certain situations or commit it themselves. The Pentagon works with a database of nearly 2,000

non-state groups, including rebel militias, complex criminal organizations, hackers, and the like. Where appropriate, they can be
deployed in the context of U.S. special operations, which subsequently are publicized at will as either 'terrorism' or 'counter-ter-
rorism.' The Pentagon also developed theories of information
warfare ('perception management') to support such operations.
Besides disrupting and demoralizing the opponent and his allies,
an important factor in this perception management is to convince
the public at home that the costs of the war effort are worth it.
In turn these programs have produced the concepts of 'network
warfare' and 'cyber war.'[61]

After the financial crash of 2008, the Obama administration
began phasing out its infantry presence in Afghanistan and Iraq,
moving to drones, mercenaries, and Joint Special Operations
Command (JSOC) deployment. Under General Stanley
McChrystal, the Commander-in-Chief of the U.S. armed forces
in Afghanistan, JSOC used electronic information to carry out
targeted killings. According to Obama adviser John Brennan,
who became CIA director in 2012, terror should be fought like a
metastasized cancer, without destroying the tissue surrounding it.
This philosophy made targeted assassination the primary activity
of the CIA.[62]

In November 2012, Obama signed an order to the Pentagon
and other branches of the government to initiate a global program
of aggressive cyber operations.[63] A month later, Glenn Greenwald
was approached by Edward Snowden, who would make this program public in great detail. The motto of the U.S. cyber operations was *Total Information Awareness:* making sure everything is
known about the population before it might even think of revolt.
In 2014, the Pentagon stepped up data collection with a program
called 'A New Information Paradigm? From Genes to "Big Data"
and Instagram to Persistent Surveillance... Implications for
National Security.'[64] Note that genetic material is listed separately
among the targets. In the next chapter I will elaborate on how the
IT revolution was shaped by the Total Information Awareness paradigm and how it was inspired by the war against the population.

In the words of Jeff Halper, 'War is thus rendered endemic, since it is neither possible nor desirable to end the "permanent emergency"... Pacifying humanity becomes the only way to remove war, but that endeavor itself becomes a violent, never-ending totalitarian project.'[65] This is what is playing out before our eyes.

The home-front application of the methods used in Afghanistan and Iraq came with JSOC commander McChrystal's move into civilian life. In a revealing report on the failure of the Afghanistan operation in *Rolling Stone* in 2010, McChrystal disparaged the political leadership in Washington; he was recalled and fired. The journalist who conducted the interview, Michael Hastings, would still publish a book about the Afghanistan adventure in 2012, but in June the year after, his Mercedes was hacked and detonated by remote control in a collision. Shortly before his death, Hastings had reached out to Wikileaks because he was working on an article about CIA Director John Brennan's role in spying on critical journalists and felt he was being followed.[66]

McChrystal, meanwhile, immediately after his dismissal in 2011 had set up an advisory group, McChrystal Group, inspired by the successes he supposedly had scored in breaking up the Al Qaeda network in Iraq. According to the McChrystal Group website, the JSOC had accumulated the equivalent of a century of military experience there and developed a team approach (combining military units, the CIA, etc.). This was now made available for domestic use. In 2020, this made the McChrystal Group the pivotal consultancy in handling the Covid challenge and it secured a steady stream of business from individual cities and states.[67] Not surprisingly, the ex-general believed that the fight against Covid-19 must be waged like a war, as he put it in an interview in *Forbes* in April. The federal government should take the lead, there being no point in having fifty states individually waging that war. Neither should there be any political intervention. The war must be waged *without opposition*, otherwise it will end, as in Vietnam, with a retreat instead of a victory.[68] No wonder that in the run-up to the November 2020 election, the McChrystal Group via its website DefeatDisinfo.com openly campaigned against Trump's handling of the 'pandemic.'

For his information war, McChrystal relies on a wide variety of allies in the mainstream media, from Fox News and CBS to CNN.[69] In addition to media representatives, the podcasts of the group, under the telling title '*No Turning Back,*' also feature spokespersons for the military-industrial complex such as Michèle Flournoy, former deputy defense secretary with the reputation as an extreme hawk, as well as representatives of the biopolitical complex such as Sue Desmond-Hellman, CEO of the Bill & Melinda Gates Foundation from 2014 to 2020. As the leading U.S. cyber-PR company, the McChrystal Group also campaigns against corona skeptics and 'anti-vaxxers,' along with the *Poynter Institute*, funded by the Gates Foundation, the Omidyar Network (of eBay owner Pierre Omidyar), Soros' Open Society Foundation, and Facebook.[70] In the next chapter I will elaborate more systematically on the broad coalition behind the Covid state of emergency.

FEAR AND COUNTERINSURGENCY IN EUROPE

The European country deemed most at risk of revolutionary threat on the eve of the Covid crisis, France, is also closest to the United States when it comes to actually deploying its military against the population. As in the U.S., the origin of the techniques applied under the Covid state of emergency can be traced back to the transition from the colonial to neocolonial subjugation of the periphery. In France, emergency rule, midway between common law and the general state of war, was instituted in 1955 to deal with the consequences of colonial warfare for the home front. After the decolonization of Algeria, the bombing campaign by reactionary military organized in the 'secret army' OAS led to the prolongation of the state of emergency in 1962, but now, with the link to foreign wars no longer immediately evident, the machinery of repression was entirely reoriented towards the domestic situation.[71]

When the West abandoned class compromise for the 'terror threat' after 9/11, France, which had still resisted the Anglo-American invasion of Iraq under Chirac, fell in line following the election of Nicolas Sarkozy in 2007. Sarkozy chose to submit to U.S.-Israeli supervision in this domain by reorganizing the French intelligence services, merging the General Intelligence (RG) and the Department of Security (DST) into a single General Directorate of Internal Security (DCRI) and placing a close associate at its head. Another Sarkozy ally was made head of the DGSE, foreign intelligence. The new DCRI intensified surveillance of Muslims and raised the level of cooperation with Israel. With his re-election in 2012 uncertain, Sarkozy was publicly advised that only a dramatic emergency might save him. Two apparently unrelated shootings took place in the south of France in March: one traced to neo-Nazis targeting North African soldiers of a parachute regiment, the other a bloody gun attack on a Jewish religious school nearby. Sarkozy threw the two together and introduced Patriot-style anti-terror laws in parliament, but they were rejected. The intelligence services then identified a single perpetrator of both attacks: a DCRI informer of Arab background, who was shot dead in a circus-like siege, unable to contradict a police résumé casting him as a murderous fanatic.[72]

Sarkozy was nevertheless defeated by François Hollande, who could not be counted on as surely as his predecessor to toe the line. Abandoning the promised end to austerity after a visit to Berlin, Hollande faced mounting popular unrest. His economy minister left in protest, to be replaced by Emmanuel Macron in 2014. Meanwhile the socialist majority in parliament voted in favor of recognizing an independent state of Palestine, and on 5 January 2015, Hollande spoke out against sanctions on Russia over Ukraine. A few days later, on the 7th, masked gunmen shot dead 12 people in the editorial office of the anti-Muslim magazine, *Charlie Hebdo*, and four hostages were killed in a Jewish supermarket shortly thereafter. As in 2012, it seemed as if an anti-Semitic attack was tacked on to an otherwise unrelated atrocity. Gilles Kepel offers an elegant explanation of 'generations' in the

Islamist jihad that would lend credence to the strategic orientation of the perpetrators, whilst explicitly rejecting the 'conspiracy theory' that undercover operatives with a background would have been involved too.[73] However, in this instance there are too many incongruities suggesting a false flag operation intended to force Hollande to change course and instill fear in French society, for one thing, and another, to encourage emigration to Israel.[74]

After another, even bloodier spate of terror attacks in Paris in November, the state of emergency was introduced and renewed five times since. The *Sentinelle* anti-terrorism operation was established in 2016. Made permanent, it has bought 10,000 troops into the streets, overseen by a Defense Council, an organ created by Sarkozy that his successors have eagerly used as a crisis tool.[75] Effectively therefore, France was placed under a *permanent* state of emergency. In facing the *gilets jaunes* in 2018 Macron made the Defense Council the central dashboard of the response; it was that body too that he convened on 4 March 2020 when the 'pandemic' was declared, making France, in the judgment of Claude Serfati, the sole Western democracy to militarize the approach to the crisis.[76] The furious attacks against demonstrators by the militarized police, against the Yellow Vests as well as in the lockdown, and the gangland practices against dissenting experts, reached a low point when the pharmaceuticals professor, Jean-Bernard Fourtillan, was arrested at his home on 10 December 2020 for his explosive revelations on the origin of the virus. He was locked up in the psychiatric hospital in Uzès, from which he was temporarily released following a nationwide wave of protest, only to be reincarcerated later.[77]

Yet even in Europe France is not the only country adopting the military approach. In Britain, a cyber-warfare unit was established by GCHQ, the UK equivalent of the NSA. It uses methods developed to combat terror networks and as in the United States, on occasion has been on the border line between fighting and committing acts of terror. Thus, as transpired in the London *Times* in November 2020, GCHQ was instructed to 'eliminate' opponents of vaccination. A secret military unit, assisted by the 77th brigade of the British army, has the remit of tracking down dissent concerning the pandemic and especially 'anti-vaxxers.'[78] Meanwhile in Germany, a special branch of the domestic intelligence service has also been tasked to track down dissidents contesting the validity of the virus narrative.

In the Netherlands, the deployment of soldiers in the information war was also stepped up in the Covid crisis. The militarization of the repression in the Netherlands built on intelligence structures that were set up after the financial crisis of 2008. These have, as their common element, the suspension of the separation of powers and its replacement by the unification of state authority against 'subversion.' The dedicated bodies established to this end, even including their own academic branch, all rely on 'data-driven operations and the development of artificial intelligence.' In September 2019, well before the 'pandemic' was declared, steps were taken to create a military center for 'information-driven action.' On 16 March 2020, the Land Information Maneuver Center (LIMC) was inaugurated to combat 'disinformation,' targeting Internet channels not already silenced by YouTube and Facebook censorship.[79] The LIMC works closely with the National Coordinator for Terrorism and Security, who has publicly identified the publisher of the Dutch version of the present book as a security risk. LIMC social media activities are organized according to the Behavioral Dynamics Methodology developed by SCL, the parent company of Cambridge Analytica, which was a key factor behind the electoral successes of nationalist populist parties, as we will see in the next chapter. To gain a handle on blocking 'disinformation,' the new designation of social criticism in the Covid state of emergency, information warfare includes engaging in discussions on social media to discover the identities of vocal critics and pass them on to the police.

While in the United States, all things military are anchored in the most comprehensive and powerful military-industrial complex in world history, Europe is catching up in this respect as well. The obvious militarization of the European Union has seen a range of EU programs being opened up to defense contractors, 'deepening the industry's involvement in previously civilian areas of European cooperation.'[80] We will later discuss the role of the EU in preparing for the 'pandemic' and the vaccination strategy well before the actual outbreak of the Covid scare. In the security field, the EU has so far left the preparations for a large-scale land war with Russia to NATO and instead has focused on complementary 'hybrid warfare' activities and the fight against 'disinformation.' The process of class formation that is driving the response to the Covid crisis in fact had its origins in the IT revolution and is focused on exploiting its achievements for population control.

Endnotes

1. Andrey Korotayev, Kira Meshcherina and Alisa Shishkina, 'A Wave of Global Sociopolitical Destabilization of the 2010s: A Quantitative Analysis,' *Democracy and Security*, 14 (4) 2018, pp. 331–57.

2. Andrew Edgecliffe-Johnson, 'Trust in governments slides as pandemic drags on: Erosion of early public support around the world threatens vaccine rollouts,' *Financial Times*, 13 January 2021 (online); *Edelman Trust Baromater 2021* [pdf] (online); Ullrich Mies in Flo Osrainik, *Das Corona-Dossier: Unter falscher Flagge gegen Freiheit, Menschenrechte und Demokratie.* (Neuenkirchen: Rubikon, 2021), pp. 18–19.

3. Joseph A. Schumpeter, *Capitalism, Socialism, and Democracy* (New York: Harper & Brothers, 1942), p. 268; on the labor market, M. Kalecki, 'Political Aspects of Full Employment' [1943], in E.K. Hunt and J.G. Schwartz, eds., *A Critique of Economic Theory* (Harmondsworth: Penguin, 1972).

4. Daniele Ganser, *NATO's Secret Armies: Operation Gladio and Terrorism in Western Europe* (London: Frank Cass, 2005), pp. 26–28; Cees Wiebes and Bert Zeeman, 'The Pentagon Negotiations March 1948: The launching of the North Atlantic Treaty,' *International Affairs* 59 (3), 1983, pp. 351–63.

5. Dietrich Rueschemeyer, Evelyne H. Stephens, and John D. Stephens, *Capitalist Development and Democracy* (Chicago: University of Chicago Press, 1992); Jean Fourastié, *Les trente glorieuses ou la Révolution invisible de 1946 à 1975* [rev. ed.] (Paris: Fayard, 1979).

6. Peter Dale Scott, *Deep Politics and the Death of JFK* [with a new preface] (Berkeley, Calif.: University of California Press, 1996 [1993]).

7. Gianfranco Sanguinetti, *Over het terrorisme en de staat* [translated from the French] (Bussum: Wereldvenster, 1982 [1979]); Ganser, *NATO's Secret Armies.*

8. Michel Crozier, Samuel P. Huntington, and Joji Watanuki, *The Crisis of Democracy: Report on the Governability of Democracies to the Trilateral Commission* (New York: New York University Press, 1975), p. 9, emphasis added.

9. See my *Global Rivalries from the Cold War to Iraq* (London: Pluto and New Delhi: Sage Vistaar, 2006), chapter 5.

10. Peter Dale Scott, *The American Deep State: Wall Street, Big Oil, and the Attack on U.S. Democracy* (Lanham, Maryland: Rowman & Littlefield, 2015), p. 149.

11. Len Colodny and Robert Gettlin, *Silent Coup: The Removal of a President* (New York: St. Martin's Press, 1992).

12. Robert Scheer, *With Enough Shovels: Reagan, Bush, and Nuclear War* (New York: Random House, 1982).

13. James Mann, *Rise of the Vulcans: The History of Bush's War Cabinet* (New York: Penguin, 2004), pp. 138–39.

14. Samuel P. Huntington, *American Politics: The Promise of Disharmony* (Cambridge, Mass.: Harvard University Press, 1981), p, 232.

15. Scott, *The American Deep State*, pp. 32, 39.

16. Glenn Greenwald, *No Place to Hide: Edward Snowden, the NSA and the Surveillance State* (London: Hamish Hamilton, 2014).

17. *Neoliberalism* refers to economic policies promoting the integral subordination of society to capital ('the market'); *neoconservatism* to a political movement dedicated to the aggressive propagation of this system across the globe by the United States, through regime change or otherwise.

18. Raúl Delgado Wise and David T. Martin, 'The political economy of global labour arbitrage,' in Van der Pijl, ed., *Handbook of the International Political Economy of Production* (Cheltenham: Edward Elgar, 2015), p. 70.

19. Werner Seppmann, *Ausgrenzung und Herrschaft: Prekarisierung als Klassenfrage* (Hamnburg: Laika Verlag, 2013).

20. Marcus Taylor and Sébastien Rioux, *Global Labour Studies* (Cambridge: Polity Press, 2018), pp. 88–89; Jeroen Merk, 'Production beyond the Horizon of Consumption: Spatial Fixes and Anti-Sweatshop Struggles in the Global Athletic Footwear Industry,' *Global Society*, 25 (1), 2011, pp. 73–95.

21. William I. Robinson, 'Global Capitalism Post-Pandemic,' *Race & Class*, 62 (2), 2020, pp. 2, 7–8.

22. Sergey Bodrunov, *Noönomy* [English version of the Russian edition], presented at the conference 'Marx in a high technology era: Globalization, capital and class' (University of Cambridge, 2018); Robinson, 'Global Capitalism Post-Pandemic,' p. 7.

23. *World Economic Forum, Resetting the Future of Work Agenda: Disruption and Renewal in a Post-Covid World* [in collaboration with Mercer] (Geneva: WEF, 2020).

24. Thomas Piketty, *Capital in the Twenty-first Century* [trans. A. Goldhammer] (Cambridge, Mass.: Harvard University Press, 2014), p. 463.

25. Jeff Halper, *War Against the People: Israel, the Palestinians and Global Pacification* (London: Pluto Press, 2015), p. 258.

26. Mike Davis, *Planet of Slums* (London: Verso, 2017 [2006]), p. 205.

27. Samuel P. Huntington, *The Clash of Civilizations and the Remaking of World Order* (London: Touchstone, 1998), p. 211. Original article: Huntington, 'The Clash of Civilizations?,' *Foreign Affairs*, 72 (3), 1993, pp. 22–49.

28. Martin Van Creveld, *The Transformation of War* (New York: The Free Press, 1991), p. 223; Halper, *War Against the People.*

29. Benjamin Netanyahu, ed., *Terrorism. How the West Can Win* (London: Weidenfeld & Nicolson, 1986), pp. 225–26; Dominick Jenkins, *The Final Frontier: America, Science, and Terror* (London: Verso, 2002), p. 75.

30. Stephen Gill, *Power and Resistance in the New World Order* (Basingstoke: Palgrave Macmillan, 2003), p. 213.

31. Claude Serfati, *Le militaire: Une histoire française* (Paris: Éditions Amsterdam, 2017), p. 203.

32. Jennifer Clapp and Eric Helleiner, 'Troubled futures? The global food crisis and the politics of agricultural derivatives regulation,' *Review of International Political Economy*, 19 (2), 2012, pp. 184–5; *Der Spiegel*, '"Land Grabbing": Foreign Investors Buy Up Third World Farmland,' 18 February 2013 (online).

33. Korotayev, Meshcherina and Shishkina, 'A Wave of Global Sociopolitical Destabilization of the 2010s,' pp. 332, 336, 338; *Edelman Trust Baromater 2021* [pdf] (online). Compare this to the preludes to the two world wars: Beverly J. Silver, 'Labour, war and world politics: Contemporary dynamics in world-historical perspective,' in Kees van der Pijl, ed., *Handbook of the International Political Economy of Production*, pp. 7–9.

34. Zbigniew Brzezinski, *Strategic Vision. America and the Crisis of Global Power* [with new postface] (New York: Basic Books, 2013 [2012]), pp. 28–32.

35. Among them, the economist, Joseph Stiglitz, and the journalist. Paul Mason. These and the ILO cited in Frank Deppe, *Autoritärer Kapitalismus: Demokratie auf dem Prüfstand* (Hamburg: VSA, 2013), pp. 268–69. See also Patrick Zylberman, 'L'avenir, "cible mouvante": Les États-Unis, le risque NRBC et la méthode des scénarios,' in Serge Morand and Muriel Figuié, eds. *Émergence des maladies infectueuses: Risques et enjeux de société* (Versailles: Eds. Quae, 2016), p. 83.

36. Brian Wheeler, 'EU should "undermine national homogeneity" says UN migration chief' [interview with Peter Sutherland], *BBC News*, 21 June 2012 (online).

37. *United Nations, Intergovernmental Conference on the Global Compact for Migration*, 10–11 December 2018, Marrakech, Morrocco (online).

38. Christophe Guilluy, *La France périphérique: Comment on a sacrifié les classes populaires* (Paris: Flammarion, 2015).

39. Ibid., pp. 26–27.

40. *Kerncijfers Amsterdam 2018* (online), p. 8.

41. Michael Lind, 'The New National American Elite,' *Tablet*, 20 January 2021 (online).

42. Jeffrey A. Tucker, 'Who Wanted Pandemic Lockdowns?,' *American Institute for Economic Affairs*, 9 February 2021 (online).

43. *Edelman Trust Barometer 2021.*

44. Guilluy, *La France périphérique*, p. 33. For example, in 2018, 456,103 people of migrant background lived in Amsterdam, of whom 265,449 were first-generation immigrants; compared to 398,113 Dutch; *Kerncijfers Amsterdam 2018.*

45. Aarti Betigiri, 'Unrest in India as it prepares for the world's biggest election,' *The Interpreter*, 13 February 2019 (online); background in Radhika Desai, 'Forward March of Hindutva Halted?,' *New Left Review* II (30), 2004, pp. 49–67.

46. *The Economic Times,* 'NCRB data shows 42,480 farmers and daily wagers committed suicide in 2019,' 1 September 2020 (online).

47. Soutik Biswas, 'Why has India's Assam erupted over an anti-Muslim" law?' *BBC*, 13 December 2019 (online); *BBC*, 'Delhi election: Arvind Kejriwal returns as chief minister after AAP victory,' 11 February 2020 (online).

48. *Worldometer*, 'current corona figures' (online), figures for April 2021. In May the figures shot up, but still a far cry from the levels registered in some other countries.

49. Sandra Cuffe, 'Chile protests 2 months on: "We're ready to continue to the very end." Government concessions have failed to satisfy protesters who've vowed to stay in the streets until their demands are met,' *Al Jazeera*, 18 December 2019 (online).

50. Osrainik, *Das Corona-Dossier*, p. 239.

51. *Wikipedia*, '2021 Chilean Constitutional Convention election.'

52. *Wikipedia*, '2019–2020 French pension reform strike.'

53. Christophe Guilluy, 'The *gilets jaunes* are unstoppable' (interview by Fraser Myers), *Spiked.com*, 11 Januariy 2019 (online).

54. Claude Serfati, 'France. "Militaro-securité globale": Le jour d'après est déjà là,' *Alencontre*, 15 December 2020 (online).

55. *Le Monde,* 'Réforme des retraites: une huitième journée de grève et de manifestations,' 29 January 2020 (online)

56. Annie Lacroix-Riz, *Les élites françaises entre 1940 et 1944: De la collaboration avec l'Allemagne à l'alliance américaine* (Paris: Armand Colin, 2016).

57. *Wikipedia*, 'French elections.'

58. Douglas Valentine, *The Phoenix Program* (New York: William Morrow & Co. [Authors Guild Backinprint.com Ed.], 2000 [1990]); Peter Dale Scott, 'The United States and the Overthrow of Sukarno, 1965–1967,' *Pacific Affairs*, 58 (2), 1985, p. 264.

59. T. David Mason and Dale A. Krane, 'The Political Economy of Death Squads: Toward A Theory of the Impact of State-Sanctioned Terror,' *International Studies Quarterly*, 33 (2), 1989, p. 178.

60. Cited in Andrew Cockburn, *Kill Chain: The Rise of the High-Tech Assassins* (New York: Henry Holt & Co., 2015), pp. 116–17.

61. Nafeez Mossadeq Ahmed, 'How the CIA made Google: Inside the secret network behind mass surveillance, endless war, and Skynet,' *Insurge Intelligence*, 22 January 2015 (online).

62. Jeremy Scahill, *Dirty Wars: The World is a Battlefield* (London: Serpent's Tail, 2013), pp. 102–10; Brennan cited in Cockburn, *Kill Chain,* p. 219.

63. Greenwald, *No Place to Hide,* p. 81.

64. Ahmed, 'How the CIA made Google.'

65. Halper, *War Against the People*, p. 29.

66. Tim Dickinson, 'Michael Hastings (Rolling Stone Contributor), Dead at 33,' *Rolling Stone*, 18 June 2013 (online); *Fox News.* Journalist Michael Hastings sent chilling email to colleagues before death' (24 June 2013) (online); *Civil Liberties Defense Center,* 'Who Killed Michael Hastings?' (13 November 2013) (online).

67. For example, the city of Boston (*MSN News*, 18 April 2020), the state of Missouri, *The Kansas City Star*, 1 September 2020, etc., etc.

68. Brook Manville, 'Are We Really Fighting Coronavirus "Like A War"?' [Interview with Stan McChrystal], *Forbes,* 2 April 2020 (online).

69. *CNN Opinion* by Stan McChrystal and Terry McAuliffe, 'How to win the fight against a virus that knows no boundaries,' 10 April 2020 (online).

70. Leonard G. Horowitz, *Complaint for Injunctive Relief Against Unfair and Deceptive Trade by Civil Conspiracy in Violation of the Florida Whistleblower Act, Civil Rights, and Public Protection Laws,* civil suit against Pfizer, Inc., Moderna Inc., Hearst Corp. and Henry Schein, Inc., in the U.S. District Court for the Middle District of Florida, 1 December 2020 [pdf], pp. 20–22, 25, 48–49.

71. Claude Serffati, *Le militaire: Une histoire française* (Paris: Éditions Amsterdam, 2017), pp. 187–88.

72. Laurent Guyénot, 'Prequel to Charlie Hebdo: The March 2012 Mohammed Merah Affair (a Mossad-DCRI Co-Production?)' in Kevin Barrett, ed., *We Are NOT Charlie Hebdo: Free Thinkers Question the French 9/11* (Lone Rock, Wisconsin: Sifting & Winnowing Books, 2015), pp. 96–101. The weekly, *L'Express*, had predicted that Sarkozy would need an 'international, exceptional or traumatizing event' or lose (ibid.).

73. Gilles Kepel, *Terreur dans l'Hexagone: Genèse du djihad français* [with Antonine Jardin] (Paris: Gallimard, 2015), p. 273 & passim.

74. The chapters in Barrett, ed., *We Are NOT Charlie Hebdo*, are very uneven but ultmately closer to what really was behind the event than Kepel.

75. Serfati, *Le militaire*, p. 190, and Serfati, 'France. "Militaro-securité globale."'

76. Serfati, 'France. "Militaro-securité globale."'

77. 'Accomplished pharma prof thrown in psych hospital after questioning official Covid narrative,' *LifeSite,* 11 December 2020 (online).

78. Osrainik, *Das Corona-Dossier*, p. 277, citing *The Times*, 9 November 2020.

79. Esther Rosenberg and Karel Berkhout, 'Hoe defensie de eigen bevolking in de gaten houdt,' *NRC-Handelsblad*, 15 November 2020.

80. Iraklis Oikonomou, 'Hijacking European Integration: EU Militarization as a Class Project,' in Kees van der Pijl, ed., *The Militarization of the European Union* (Newcastle: Cambridge Scholars Publishing, 2021), p. 32.

3.

RESTRUCTURING THE RULING CLASS IN THE I.T. REVOLUTION

Responding to challenges to the existing order, especially when it comes to unrest from below, is by definition the task of the ruling class. In the West, this is an Atlantic ruling class, formed in the course of the British colonization of North America. Despite the war of independence by the United States, the common language and political traditions remained and following the victory of the industrial North in the Civil War, an influx of British capital for railway construction engendered new interconnections. The large corporations formed in the same period in turn expanded abroad, and the world wars gave the United States an enduring presence in Western Europe. Thus, an Atlantic *political* system was created too, in which major changes occur synchronously. And accordingly, either the whole system is imbued with a new dynamism and expands, as for instance in the Kennedy and Reagan years; or it loses ground and is plagued by internal fractures, as happened in the periods in between and since 2008.[1]

The Covid state of emergency is an attempt to turn the current tide of decay and disintegration and exploit the opportunities created by the IT revolution, if only to avert the danger of a democratic alternative. In this chapter we discuss how the privatization of new technologies in the defense and intelligence spheres gave rise to the large IT monopolies. The national security and intelligence sectors, the Internet and related concerns, and the (multi-)media conglomerates in combination constitute a triangle at the core of the power bloc behind the 'new normal' that is being imposed in the current crisis. In itself, there is nothing new about such a transformation: modern capitalism has reinvented itself a number of times around a power bloc, from which emanates the informal quasi-government program that allows the ruling class to stay on top for a certain period. Such a program, or comprehensive concept of control, combines the priority interests of the central power bloc with those of successive allies and crucially, manifests a widely shared sense that it represents the most appropriate approach to the problems of the day.

In the current crisis, however, the ascendant power bloc has preferred not to wait for the formation of a such a broader coalition; the signs of a revolution are too serious. Instead, it seeks to impose a digital surveillance society from above, with an appeal to the public to acquiesce because of the 'pandemic.' At the same time, the financial sector is being reorganized to avoid another collapse like the one in 2008. National populism, finally, far from representing a revolutionary force, basically represents the interests of the ruling class, albeit of a segment or a *fraction* of same. Today, unlike the situation in the 1920s and 30s. the political center constitutes the paramount threat to democracy. National populism may be useful as a force for breaking up the solidarity among the lower classes, but for the new normal it is an obstacle, as was most evident in the center's campaign against the re-election of Donald Trump.

THE I.T.-MEDIA POWER BLOC AND THE INTEGRATED COMPREHENSIVE SURVEILLANCE SOCIETY

The technological component of the information war dates back to World War II, when Great Britain and the United States joined forces to eavesdrop on telephone and other communications signals. In 1947–48 the Five Eyes was created when the British brought Canada, Australia and New Zealand on board. This English-speaking intelligence network continues to be an important support structure for the Atlantic ruling class, working closely with the services of vassal states such as Germany and France, South Korea and Japan, as well as ally Israel.[2] During the war, the U.S. also combined the first computers into a system of integrated air defense, whilst developing related systems theories such as cybernetics.[3] The British contribution was *operations analysis* for anti-submarine and air warfare, in which vast masses of variables can be arranged by a computer using algorithms.[4]

The IT revolution received a major boost when the Soviet Union in 1957 surprised the world by launching the first man-made space satellite, Sputnik. The U.S. responded within a month with the establishment of the Advanced Research Projects Agency (ARPA) that later attached the prefix 'Defense.' DARPA, as I will label it hereinafter, became the center of the IT revolution in the United States, applying inventions such as the integrated circuit and later, the microprocessor. Within three years after Sputnik, a prototype of the Internet had already been developed for air defense purposes at MIT, and in 1969 the first computer link was operational between two California universities. In 1972, an NBC correspondent made public that there existed a computer network sharing data about political opponents for the benefit of the CIA and the NSA. Preventing social revolution was the central purpose of the new information technology from the beginning.[5]

In the wake of the Sputnik success, the USSR also began to set up digital networks for its planned economy, as we will see in Chapter 7. However, after the fall of Khrushchev, the initiative shifted to the West again. When the Nixon administration suspended the dollar's gold backing in 1971, vast public funds were invested in the development of digital communications, computers, and the like, as the need to balance government expenditure with tax revenues had gone overboard. The value of the dollar henceforth depended on the confidence of the global ruling classes in American 'leadership' and on the extent to which the dollar was kept in use as a monetary reserve and as means of payment for raw materials. Thus, by taking advantage of tax freedom and other benefits, Silicon Valley emerged. No other capitalist country enjoyed this fiscal luxury or had such a powerful military apparatus (although the USSR did try to keep up at the expense of its own civilian industry).[6]

The hippy subculture of Silicon Valley, from which the personal computer originated, should not obscure the fact that the first Apple in 1976 also used technologies funded by DARPA, that is, by the Pentagon. A year later, the prototype of an Internet connection was used to conduct a virtual military exercise with England and Sweden via satellite. Shortly afterwards, DARPA started subsidizing research at Stanford into artificial intelligence (algorithms for search engines).[7] In the 1980s, it came to light that Reagan's national security adviser, Admiral John Poindexter, had drawn up a directive to have all computer files in the U.S. checked by the NSA—leading to his forced resignation. Hibernating in a company that worked for DARPA, he later became head of the Total Information Awareness agency within that institution. This agency focused on 'behavioral profiling,' 'automated detection, identification and tracing' and other data collection projects.[8] Originally these were intended to track down 'terrorists' and to justify the suspension of civil freedoms; something for which 'the virus' is being used today.

In the 1990s, when capitalism no longer tolerated political direction following the collapse of the Soviet bloc, the achievements of the IT revolution fell into private hands. In 1995, the National Science Foundation handed over NSFNET, the rump-Internet, to a group of private providers. A year earlier, the first search engine had been developed by the Digital Library Initiative (a program of the National Science Foundation, NASA and DARPA); the same program also funded two Stanford Ph.D. students, Sergey Brin and Larry Page, to develop an automated search engine based on the frequency of citations. In addition, they set out to develop a method to adapt the information thus retrieved to the interests of the user, which would become the core of artificial intelligence.[9]

After Brin and Page took their private company, Google, to the Nasdaq technology stock exchange in 2004 and became billionaires, ties with the national security state were not severed. Metadata is collected via Google's gmail to obtain a complete profile of the user; Google also has a separate director who is in constant contact with the National Security Council in Washington and the intelligence agencies. That these data are being collected for commercial purposes is only half the story.[10]

The other IT giants, such as Apple and Sun Microsystems, also have a background in defense intelligence. Hardware too was often a product of military research, e.g. the LCD screen, the touch screen, etc.[11] Through contracts with the CIA, the NSA and the National Geospatial Agency (NGA), which operates space satellites, the IT groups remained part of the U.S. defense and intelligence system. In turn, defense infrastructure was made available for general use by commercializing certain technologies, such as the Keyhole satellite surveillance program that spawned Google Earth.[12] The key concept is always Total Information Awareness because 'private' use is also monitored and ultimately ends up in national security databases too.

I.T. AND THE SHADOW BANKING SECTOR

The IT sector had gained in luster partly due to the application of specific technologies in the liberalized financial sector ('FinTech') and the record profits made there. The Banking Act in 1980 made money creation outside the regulatory purview of the U.S. monetary authorities possible; non-bank institutions henceforth were allowed to create money through credit and provide financial services. This called into being an entirely new category of players, the so-called shadow banks.[13] The collapse of the Soviet Union shifted the perspective of money traders in the direction of pure arbitrage and commercial profit. 'Chasing yield' replaced any long-term view and not only market operations, but politics too, was transformed into a matter of risk taking and gambling.[14]

The exponential expansion of financial transactions also led to the data revolution in IT, Big Data. Already in the 1980s individual computers were no longer able to store all the data generated by financial innovations, which required parallel data storage. The discount chain, Kmart, which had begun offering financial services to private account holders, e.g. money market certificates such as its own Kmart Fund, acquired the first parallel computer data storage system from Teradata in 1986.[15] In 2007, the U.S. shadow banking sector, with $13 trillion in assets under management, was one-third larger than the regulated banking sector ($10 trillion). Going forward, the exploding financial sector developed in parallel with the IT industry, including offering online purchases of goods and services, and to develop further became dependent on IT innovations to an ever-growing degree. Private data were being collected on an unprecedented scale in order to make decisions on consumer credit, for instance.[16] The shadow banks (hedge funds, investment banks, pension funds, and so on) were pioneers in developing the new financial instruments that would eventually cause the collapse of 2008.[17]

That is precisely why the financial sector is *not* at the heart of the bloc that has seized power under the Covid state of emergency. Certainly not the speculative capital that came to predominate in the two decades preceding the crash. True, the profits of the shadow banks recovered, but as a capital fraction this sector is no longer able, if it ever was, to provide the strategic direction to a broader class coalition, nor can it offer the population anything other than austerity, whilst any resistance will be met with physical violence. The latter has so far remained marginal, reserved for protest demonstrations against 'the measures,' because the façade of democracy must remain intact until the very end, even during the seizure of power. An unbridled assault on the population carries great risks and can even lead to a general political collapse. The possibility that the police may refuse to continue to attack their fellow citizens is such a risk.

Later in this chapter I will come back to what then happens with the financial sector instead.

INFORMATION POWER: THE INTELLIGENCE-I.T.-MEDIA TRIANGLE

The regrouping of the ruling class in the IT revolution obviously has its pivot in the IT sector, but to understand how this works in practice, we have to go back to the challenge that the existing order faces *as a system*. That is the threat of a new '1848,' a general popular revolt. In other words, there has been an objective shift in the relations of force that also requires an objective response and under the given circumstances, it is the IT revolution that offers the answer—again, at the level of the system, of the property relations in which the power of a ruling class is anchored.

The IT revolution has provided the private groups and the oligarchs behind them with the objective ability to evade a physical war against the population via the alternative of permanent surveillance and information warfare. This alternative, based on Total Information Awareness, evokes the image of the dome prison in which the cell blocks are grouped in a circular pattern, enabling them to be permanently watched from a central observatory, the *panopticon.* The design for such a prison was made by the British thinker, Jeremy Bentham. In the '30s of the 19th century Bentham exchanged the optimism of the Enlightenment and the French Revolution for a more restricted view: it was enough that the greatest possible number of people would achieve happiness, the rest simply had to be kept under control or even locked up. Several authors have recognized that this panopticon is the model for our form of late capitalism, the surveillance society organized and operated by the IT industry.[18]

While the monitoring of financial transactions will always remain incomplete due to the many offshore opportunities, tax havens, etc., the monitoring of the population can in principle be made watertight now that life is increasingly mediated through the Internet. In 2019 overall Internet traffic was sixty times greater than in 2005, while the global Internet Protocol (IP) traffic, a proxy for data, grew from about 100 gigabytes (GB) per day in 1992 to more than 45,000 GB per second in 2017. And yet the world is still only in the early days of being data driven; by 2022 it is expected that global IP traffic will reach 150,700 GB per second, fueled by more and more people coming online and the expansion of the Internet of Things, machines connected to the Internet.[19] In addition, methods of tracing the physical movement of individuals and their contacts are now being developed, for which the 'pandemic' provides the pretext. The 'contact app' on one's mobile phone following 'vaccination against Covid' is but one step to make that happen.

The large IT concerns—Microsoft, Apple, Amazon, Facebook, and Google (in China, Tencent and Alibaba), to which Zoom and similar companies have now been added—are perfectly suited for the abovementioned range of tasks. They have developed the instruments for it themselves and have also experienced enormous financial growth in the past decade. Apple and Microsoft had a market capitalization of $1.4 trillion each in 2020, followed by Amazon at $1.04 trillion, Alphabet (Google's parent company), $1.03 trillion, and Facebook, $604 billion. The capitalization of Samsung (South Korea) stood at $983 billion, Alibaba's and Tencent's at around $500 billion each. Since Google was worth less than $200 billion in 2008, we are talking about a fivefold increase. In total, the market value of the IT groups with a market capitalization of more than $100 million rose by 67 percent to more than $7 trillion.[20]

Their class formation as a capital fraction to exercise organized social and political power followed. In 2017, 2030Vision was set up at the initiative of ARM, the British chipmaker owned by Japan's Softbank. Hosted by the World Economic Forum and endorsed by the UN Secretary-General, its current members are Amazon, Google, Facebook, Salesforce, Hewlett-Packard, Unilever, McKinsey, Huawei, several UN agencies, and the government of Botswana (!). They all view 2030Vision as the model for the global public-private partnership that is going to tackle key challenges, this time with the help of new technologies.[21]

The wealth of the billionaires associated with these IT corporations increased by another $1,064.3 billion, more than a third, to $4,011.8 billion in the course of the Covid crisis.[22] The top 10

U.S. billionaires together are now worth more than a trillion. How does this contrast with the dramatic impoverishment of the rest of the population? In the U.S. alone, there are 67 million people out of work, 20 million relying on benefits, 98,000 company closures, no more medical insurance, not enough food, etc. That is another world. Here our concern is the formation of a new power bloc that has the will and the resources to discipline a world population that has become restless, and secondly, as far as U.S. and other Western billionaires are concerned, the desire to confront the Chinese challenge. The extreme concentration of wealth in the hands of this oligarchy also means that their personal idiosyncrasies have begun to exert an unprecedented, disproportional influence on social life.

Amazon's Jeff Bezos, who owns the *Washington Post* and is associated with the CIA to whom he has leased part of the Amazon cloud for Big Data storage, earned $71.4 billion in the crisis and climbed to $184.4 billion; the earnings of his ex-wife, MacKenzie Scott, climbed more than $20 billion to nearly 60. Facebook's Mark Zuckerberg doubled his assets to more than $100 billion during the first nine months of the crisis. Brin and Page, the founders of Google, each cashed in about $28 billion extra in the period. And so it goes. There are also billionaires who have become richer in other domains, most spectacularly Elon Musk of Tesla (electric cars) and SpaceX (satellites), who earned $118.5 billion in the crisis to climb from $24 billion to $143.1 billion and became the second richest man in the U.S. after Bezos. However, the IT capitalists are leading the way.

Bill Gates's earnings climbed by $20 billion to $118.7 billion. Although Microsoft's contribution to the technical development of both the computer and the Internet is negligible (compared to Oracle, Apple, Sun Microsystems, Intel, and others), Gates, by his clever marketing strategy, succeeded in making MSDOS and then Windows the world standard. Regular 'updates' of these operating systems feed the suspicion that the intelligence services all along had a back door in these systems allowing them to inspect the content of individual laptops, plant cookies, and the like. The link with the national security state after all has only grown closer: Gates has made the JEDI cloud (value $10 billion) available to the Pentagon and, as mentioned, the Amazon cloud contracts with the CIA, which other intelligence services can also use. This is part of the development of a world-wide system of digital interfaces, in which the pharmaceutical and biotech sectors too have become partners, as we will see in Chapter 6.[23]

Here the Gates Foundation, which operates mainly in the health sector, plays a key role. By setting up his own tax-exempt foundation Bill Gates followed in the footsteps of the famous robber barons in the late 1800s, early 1900s. They all converted their wealth into social power in this way: the Rockefeller Foundation, the Carnegie institutions, and later the Ford Foundation, were long the largest and most influential. Today the Bill & Melinda Gates Foundation is by far the richest of them all (the second in terms of capital invested is the UK's Wellcome Trust), and few are run so autocratically. The Gates Foundation has no Board of Trustees and Bill and Melinda shared power only with Bill's father William H. Gates Sr. (who passed away in September 2020; the divorce of the Gates may bring further changes). The various programs of the Gates Foundation have their own directors, but the program presidents, the investment managers, and the CEO of the Foundation were always directly appointed by Bill and Melinda. In addition, the Gates Foundation is primarily a tax-free investment outfit; for every dollar in donations, it has channeled a multiple into large companies, both pharmaceutical companies (of which more later) as for example, BP and ExxonMobil.[24]

Gates, then, is not an IT revolutionary like Steve Jobs of Apple, but above all a class-conscious strategist. He sees his own role and that of his foundation as a contribution to 'creative capitalism,' supported by 'catalytic philanthropy.' The Gates Foundation is expected to 'use all the tools of capitalism to link the promise of philanthropy with the power of the private enterprise,' according to Gates.[25] All leaders, both in politics and in business, must of course have some sort of ruler's instinct, and Gates is certainly no exception. Early on, he was referred to as 'Citizen Gates' (a variant on the megalomaniac newspaper mogul *Citizen Kane* in Orson Welles' eponymous film), who wanted to take over not just the software industry but the entire world—on the basis of the draconian terms of employment of Microsoft staff, whom he wanted to be kept on temporary contracts to get the most of them.[26]

Gates also plays an important role in the Democratic Party, which has increasingly become the political tool of the oligarchy and the intelligence-IT-media bloc in the U.S. In 2008, he supported Obama's candidacy, along with other members of the New America Foundation (Google and its former CEO, Eric Schmidt, the Ford Foundation, etc.). When Obama's re-election in 2012 was jeopardized by the U.S. Supreme Court ruling in the *Citizens United* case that corporations have the same right to free speech as individual citizens, and a host of far-right billionaires set up Super-PACs for their own candidates, Obama was able to call on Gates to mobilize his own Super-PAC to salvage his campaign.[27]

At the same time, Obama's continuation of the wars in Central Asia, the Middle East, and North Africa pushed the party further in the direction of interventionism. The Neocons were initially mainly based in the Republican Party (from Reagan to George W. Bush), but after Trump's 2016 victory, made their home in the Democratic Party. This was reflected in a regrouping of thinks tanks such as the formation of the Alliance for Securing Democracy (against Russia and China), an initiative under the umbrella of the German Marshall Fund, co-directed by the current Biden National Security Council official, Laura Rosenberger, and Zack Cooper of the American Enterprise Institute. As elsewhere, prior 'left/right' divisions were being blurred by the broadening center, as in the case of the formation of the Quincy Institute by the Neocon billionaire Charles Koch and the neoliberal, George Soros, although in this case the convergence was motivated by the concern to stop foreign wars, not have more of them.[28]

Piketty's statement that we are moving towards a world in which every country is owned by its own billionaires applies, especially in the U.S., to the IT oligarchs.[29] In addition to the lasting connection with the military intelligence sector from which the Internet giants emerged, (multi-) media companies too were linked to this complex in the 1990s. Under Clinton the Telecommunications Act of 1996 gave free rein to the merger between cable operators, radio, film, and newspaper corporations, telephone companies and TV stations, as well as Internet providers. Thus, in combination with the IT giants and the intelligence world looming large behind them, an apparatus for information warfare was being created the likes of which the world has never seen.[30] An intelligence-IT-media bloc, moreover, in the hands of a very small group of people.

Due to mergers under the Telecommunications Act, of the fifty or so companies which divided the U.S. media market among themselves in the early 1980s, *six* have been left. These Atlantic media giants, in combination with Anglophone cultural hegemony, to a large extent control global information flows. The Homeland Security Act explicitly mandates the realization of a comprehensive national plan for information provision on all channels in order to ensure information flows synchronized with American policies.

The six largest multi-media companies today are Comcast (Ralph Roberts family, with MSNBC and others); Disney (including ABC); TimeWarner (including CNN); 21st Century Fox (including Fox News, whose CEO Rupert Murdoch, via News Corp. owns *The Wall Street Journal,* the *New York Post,* and in the UK, *The Times*, *The Sun*, Sky TV...); the German Bertelsmann group of the Mohn family, operates the RTL TV network and a series of top publishers (Penguin, Simon & Schuster...). Finally, there's Viacom with CBS, of the Redstone family. In the media groups that are not entirely family-owned, the three largest passive index funds (which we will discuss later) and the main banks control large blocks of shares.[31]

Bill Gates does not have a media empire of his own, although MSNBC started as a partnership between Microsoft and NBC, then still owned by General Electric. There is, however, a wide range of media that he generously subsidizes. Tim Schwab investigated which media outlets receive Gates grants and found that $250 million in subsidies was distributed to the following organizations. First and foremost, the BBC (one-fifth of the total), followed by NBC, Al Jazeera, ProPublica, *National Journal*, *The Guardian*, Univision, Medium, the (now Japanese-owned) *Financial Times*, *The Atlantic*, the *Texas Tribune*, Gannett, *Washington Monthly*, *Le Monde*, and the Center for Investigative Reporting. In late 2018, the German weekly, *Der Spiegel*, received $2.5 million from Gates to write about 'Global Health and Development.' Several of these media organizations are already in the hands of large financiers who are actively involved in editorial policy, but Gates can still add accents in this or that direction. Given his central role in the Covid crisis and especially the 'vaccination' route that he believes must be taken at all costs, this is an unprecedented outreach.[32]

Michael Bloomberg, originally a banker at Salomon Brothers before making his fortune with an eponymous media company focused on financial news (his net worth stood at $55 billion in 2020), former mayor of New York and presidential candidate, is playing his own part in the Covid panic through his record donations to medical institutes at Johns Hopkins University. Why the global media erupted about 'an unknown virus' around 20 January 2020, (and then did not stop), may be traced back to a report from the Bloomberg-funded Johns Hopkins Center for Health Security, although other connections were also at play. Johns Hopkins began issuing daily updates on the spread of 'the virus' from that time on. The extreme concentration of media capital and the presence of several owners at the WEF in Davos that same week (20–24 January) will certainly have facilitated the simultaneous release in so many different outlets. After all, the 'pandemic' itself was hardly significant at that time (on 1 February, there were fewer than 200 recorded 'cases' outside China).[33]

Media and PR agencies, like the IT sector, are also linked back to the military and intelligence complex. Many journalists are on the payroll of the secret services (the German journalist, Udo Ulfkotte, made spectacular revelations about this) or are linked to the NATO think tanks, the Atlantic Council, the German Marshall Fund, and others. The CIA invests in a PR firm, Visible

Technologies, which tracks more than half a million social media posts every day for 'Open Source Intelligence,' from Twitter to small web sites. A year after this company was founded in 2005, it partnered with WPP, the largest PR conglomerate in the world, comprising 125 separate PR and marketing firms, with famous names such as J. Walter Thompson, Young & Rubicam, Burson-Marsteller, Ogilvy, etc. etc.[34]

A parallel information warfare structure, the so-called Integrity Initiative, was created by the British Ministry of Defence. It was launched in 2015 by the London-based Institute for Statecraft and is entirely dedicated to anti-Russia propaganda.[35] The Integrity Initiative works closely with the Public Diplomacy Division at NATO headquarters in Brussels and disseminates information through think tanks such as Egmont, Chatham House, and others. Various websites operate under the patronage of the Integrity Initiative, including Buzzfeed, Irex, Detector Media, and Bellingcat.[36] Bellingcat, which is also linked to the Atlantic Council and King's College London, has played a key role in spreading the NATO account of the downing of flight MH17 over eastern Ukraine, the Western reading of chemical weapons incidents in Syria, and other propaganda stories serving the Atlantic line against its enemies. Its headquarters are in the Netherlands where it has meanwhile become interlocked with a series of other media and propaganda outlets and semi-official institutions such as Leiden University.[37]

From the foregoing, I conclude that the central role in developing the response to the restlessness of the global population has fallen to the triangle of information warfare as pictured below:

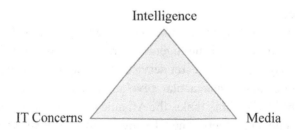

Intelligence

IT Concerns   Media

This is the core around which the ruling class in the West began to regroup after 2008 and which is now waging the information war against the global population by means of the Covid state of emergency.

THE 'SCREEN NEW DEAL'

At the heart of the plans made by the intelligence-IT-media complex to restore order (whilst advancing Western power, maximizing profits for the owners, and other individual, subsidiary motives) is the broad strategy of digitizing all of humanity with a view to bringing about an integrated surveillance constellation encompassing the entire world population, referred to above as the panopticon. Eric Schmidt, the influential ex-CEO of Google, told a video conference organized by New York State Governor Mario Cuomo in May 2020 that the new social model should be based on replacing face-to-face activities by digital interfaces. Distance learning, '*telehealth*,' i.e. health care via the Internet,[38] and all other retail too by way of online orders, are the main components of this grand scheme. Naomi Klein calls this the 'Screen New Deal,' for which Cuomo had already enlisted the help of Bill Gates, whom he called a 'visionary.' We need to implement Gates's ideas, Cuomo maintained, because why fund all those buildings and physical classrooms if we can move straightaway into the beckoning future of the digital world?[39]

However, the plans of Gates, Schmidt and other leading minds go considerably further than just distance learning and mail-ordering. The entire society must be divided into IP addresses monitored round-the-clock; after the contact app, the next step may be actual remote control. As we see in Chapter 6, research is already being carried out on behalf of the Gates Foundation to develop the implantation of nano-beacons in humans in order to use their bodies for commercial purposes. This is not limited to pharmaceutical applications either. At the beginning of 2020, Microsoft took out a patent on the use of the human body as an energy source for the creation of cryptocurrencies ('*mining*').[40] I will return to this later. Both Gates and Jeff Bezos have also entered the health market, which boasts U.S. $3.5 trillion annual sales in the United States alone; Google has set up its own medical research institute, Verily Life Sciences, through the Alphabet holding company, together with the universities of North Carolina and Harvard.[41]

Moderna, one of the producers of the revolutionary gene therapy that is now massively used as a vaccine in the Covid crisis under a provisional approval, explains that 'an mRNA technology platform ... functions very much like an operating system on a computer. It is designed so that it can plug and play interchangeably with different programs. In our case, the "program" or "app" is our mRNA drug.'[42] In other words, we are no longer speaking about ensuring public health, but rather about gaining access to a biomass of 7.5 billion humans which, as with computers, can be updated and monitored in this way. This mass will have to be mapped out, and as we will see later, people will then need bio-identity documents, in which all the major IT concerns have meanwhile invested. Catherine Austin Fitts considers Microsoft to be one of the main links through which a digital payment system combined with biometric and tracking systems is likely to be introduced, ultimately via injections.[43]

Gates, Schmidt and Bloomberg have formed a committee at the invitation of Governor Cuomo to make plans to outsource public schools, hospitals, police and other public services to private technology companies. As the head of a Maryland self-parking company puts it, the pandemic reminds us that humans are biological risks, but machines are not. Schmidt has already held various previous positions enabling him to promote his ideas, such as the Defense Innovation Board, which advises the Pentagon on the application of artificial intelligence in the defense sector, and the National Security Commission on Artificial Intelligence (NSCAI), which advises Congress and which Schmidt chairs. The Defense Innovation Board is mainly composed of academic IT specialists; the NSCAI includes the CEOs of major tech companies, In-Q-Tel (the CIA's venture capital arm), and others. At NSCAI meetings in 2019, the alarm was raised about the lead taken by China in the field of digital surveillance (thanks in part to 5G) and mobile phone payments systems; Alibaba, Baidu and Huawei were reaping the profits. Schmidt warned that thanks to artificial intelligence, China would be well past the United States by 2030. The West's response should include creating 'smart cities,' but a Google project to test this out in Toronto had to be discontinued due to popular opposition to the permanent surveillance that was inherent in it.[44] As this indicates, the idea of a panopticon is not a wild fantasy but is already being tried out under all kinds of different names.

All along, the 'triangle' advances as a single complex, with continuous feedback between its different components. In the U.S., because of the taboo on state aid, all innovation is preferably directed along the defense budget.[45] Globally, total defense spending, excluding secret budgets, increased 50 percent between 2006 and 2015 to $2.03 trillion. Digital applications for defense purposes, including actual warfare, social control, and repression, represent a market of hundreds of billions. The global market for biometrics alone was expected to more than double to $35 billion between 2015 and 2020, and close cooperation between governments and businesses is the key to this, following the Chinese example. In the West, this is easiest to achieve through the privatization of public services, so-called public-private partnerships.[46] Throughout, the war propaganda in the mainstream media is being maintained at a level never before seen, in what, for the general public, is peacetime.

Before going into the details, we must first recapitulate how a power bloc such as the triangle of intelligence services, IT concerns and media, functions as a pivot within the ruling class that propagates a particular ideology through which power can be consolidated, as is happening now with the Covid state of emergency. It should be noted immediately that the IT oligarchy need not resort to heavy-handed coup techniques, for it can count on widespread enthusiasm for everything related to the new communication technology—not only among the capitalists directly involved, the money traders and the like, but by people in all walks of life. The tech sector enjoys enormous goodwill among the public, and especially among the young who are growing up with the Internet. IT is associated with progress, increased personal power, limitless possibilities, etc. In the final chapter we will see that this in itself is a correct estimate, although the main task of the ruling class in the current crisis is to ensure that this does not fall victim to a democratic takeover and that before anything like a progressive social transformation can occur, we will all be vaccinated.

CONCEPTS OF CONTROL AND CLASS FORMATION

The Covid health crisis and the state of emergency which has been applied to it in all openness, is intended to bring about a 'new normal,' a different social order albeit one *within* the existing system of capitalist property relations.[47] Clearly the apparent success of the Chinese model has not gone unnoticed.

While the shock at how this is being introduced has no precedent in peacetime, the idea of a new normal is not as new as it sounds. Political science, the science of the exercise of power, has always been concerned with the question of how a minority can make the vast majority of the population agree to its rule, and it has never succeeded in continuing to do this on one and the same basis. One of the founders of modern political science, Gaetano Mosca, argued in his *Elements of Political Science* in 1896 that the ruling class can maintain its power by relying on a middle-class cadre of technical and administrative specialists. This cadre fashions a particular *political formula.* Such a formula, which is regularly adapted to changing circumstances, ultimately rests on an organic, historically grown unity, e.g. that of a religion, a civilization, or of the nation itself, which unites the individuals.[48]

The ability of these bonds to convince a population that the rule of a minority is normal was subject to serious erosion in the first half of the twentieth century, however. Fascism marked a final exaltation of nationality in order to stave off the danger of social-ist revolution. After its defeat, Christian democracy emerged in Western Europe to politically exploit what remained of religious solidarity. But in the English-speaking countries, a more busi-nesslike, calculating approach came to prevail over the old and established social bonds. As a result, political power increasingly came to rest on a rational social contract, essentially a class com-promise—starting with the New Deal in the U.S., which was then largely extrapolated to Western Europe as well. Japan and South Korea developed their own variants with a greater emphasis on traditional authoritarian structures carried over from the contender state experience.[49]

A political formula now has its basis in what Ries Bode calls a 'comprehensive concept of control,' which can be traced back to a particular segment, or fraction, of the capitalist class (Bode also makes a distinction between a fraction of capital and a fraction of the bourgeoisie, but that need not concern us here).[50] Such a fraction acts as an organizer of a large bloc of forces, or historic bloc, and propagates a certain notion of the public interest from the vantage point of its own interest. In the present context, this di-rective fraction would be the intelligence-IT-media triangle. Later in this book, we will see that other capital fractions, notably the biopolitical complex consisting of the health and pharmaceutical industries and the private and public bodies associated with them, have aligned themselves with this vanguard.[51] Their role is to pro-vide both the pretext and a key element in the implementation of the panopticon, for whereas a smartphone may get lost or be used by others, the human body itself cannot, and inoculation provides the route into it.

The balance of forces around the Second World War was clearly different from that of the present, and accordingly there was a different vanguard arising from it. Corporate liberalism (a liberalism of large organizations) had the new mass production industries at its center, while the financial sector, held responsible for the Great Depression and all that it entailed, was forced to lim-it itself to lending to companies. In 1936 Keynes even advocated a 'euthanasia' of the rentier, the 'functionless investor.'[52] Organized labor, on the other hand, was granted a junior role in the historic bloc, provided it went along with anti-communism and the arms race of the Cold War, as well as colonial and neo-colonial adventures in the Third World.

When progressive forces became too strong in the late 1960s, capital rescinded the post-war social contract. With the aforementioned abandonment of the U.S. dollar's gold backing and the end of fixed exchange rates, international financial markets revived. In hindsight it was not a euthanasia, but a long hibernation in the City of London that had happened to the international financiers, and they now woke up from it with a vengeance.[53] Hence the concept of control that took shape in the 1970s and 1980s—neoliberalism, the free market ideology of Hayek and Friedman—is traceable to the rentier view of the world. No longer marginalized, the financial fraction became the hub of the new historic bloc, and it was this 'new normal' that took the West to the brink of collapse in 2008.

FROM SOCIAL CONTRACT TO WORST-CASE SCENARIOS

The Lockean, liberal heartland in which capitalism and the Atlantic ruling class developed, had already entered an existential crisis in the 1990s, despite the triumphant proclamation of *The End of History*.[54] The system lost its inner coherence now that the adversary that had enabled political unity in the West had collapsed. Because the prop of anti-communism had fallen from the post-war consensus, the 'victory' in the Cold War actually exposed a moral and ideological void in which the question arose: What next? Tony Blair, in whom progressive England had still invested all its hopes declared at the Labour Party conference in 1995, a year before his election as Prime Minister, that 'we enjoy a thousand material advantages over any previous generation and yet we suffer a depth of insecurity and spiritual doubt they never knew.'[55]

With a population in that state of mind, a concept of control based on a social contract around a certain, well-functioning variant of capitalism, can no longer come about. Yet consensus is a necessary condition if a majority is going to accept being ruled by a minority. We therefore see that in the years after 1991, conscious efforts were being made to develop new ideological schemes, *scenarios*, now based on fear. In these scenarios, imaginary dangers are magnified into disasters that hang over all of us ('worst case scenarios'). While the material benefits that previously had been the reward for political conformity may have disappeared, fear remains a seemingly inexhaustible reservoir for bringing people into line. With the help of these worst case scenarios, what Patrick Zylberman calls a 'world market of fantasy' was created, because in the same period, the 24-hour news channels emerged and the Internet emerged as a global hub for unlimited information.[56] In other words, the intelligence-IT-media complex was moving towards the center of power, but the 'world market of fantasy' was in many ways uncharted territory in terms of the concrete exercise of authority.

Huntington's thesis of a 'clash of civilizations' of 1993, may be regarded as one such fantasy scenario.[57] This theory had great influence and in the changed circumstances helped to reactivate the fear of terrorism, which had initially been launched from Israel, with 'Moscow' as the center. An even more emphatic design for the direction of public opinion was discussed at a conference at the Miller Center of the University of Virginia in October 1998. There, Miller Center director Philip Zelikow outlined how politics in each era of roughly a generation is constructed around certain public myths. These myths are sets of ideas that the masses assume to be true (even if they are not entirely sure) and that are shared within the 'relevant political community,' which actively propagates them. Here we see again a reflection of Mosca's original idea, namely that a middle class must become a conduit for a political formula, or concept of control. For the generation of World War II, Zelikow argued, 'Munich' was such an idea (giving in to a dictator); as a public myth it could be invoked in the Cold War and again today.

This is not the same as a political formula, which is objective (nation, religion, civilization are not invented categories), nor is it the same as a concept of control based on a rational exchange (economic well-being for accepting imperialist politics and rejecting socialism). Now we have entered the era of scenarios and they have been precisely invented; they are largely products of fantasy.[58] But how to get them accepted if not by an organic class compromise? Here Zelikow takes another look at history, but through a new lens. Certain formative events, he argues, continue to exert their influence long after the generation that experienced them. Pearl Harbor would be such an example. Of course, Zelikow's later role in the follow up to such 'shocking events with a formative effect' is itself highly significant. Not only had he predicted the 9/11 attacks with uncanny precision; besides serving as a member of the transition team for the newly 'elected' president, George W. Bush, he also edited the final report on 9/11, which may well be considered the blueprint for a cover-up strategy.[59] There is no doubt that this scenario created a working public myth, with a shocking, formative event serving to catalyze public fear.

Clearly a scenario such as the attacks on 9/11 and the subsequent War on Terror (intended to get the United States to henceforth fight Israel's wars for it),[60] was aimed at creating fear, indeed panic. With 'terrorism' made the regulator of the political process, a population that can no longer expect anything positive from capitalism as such, can still be held under its spell. This is exactly what is now being achieved with 'the virus.' Giorgio Agamben in fact sees this as a substitute for religion now that the economy no longer 'delivers.' After having replaced the politics of religion and nation by the economy, 'biosafety' has been introduced as the new religion and so far, has achieved unprecedented success.[61]

What is new about Zelikow's approach is that instead of addressing an organic historical unity, as it resounds in a political formula, or an economic-rational synthesis that takes shape in a concept of control, he suggests a formative, shocking event can serve as the binding factor. Implicit here is the assumption that, should same not be forthcoming, it could be staged or brought about in some other way, be it via a terror attack or a 'pandemic' like the one we are now experiencing.

This is very different from reliance on the organic, moral authority that is fundamental to previous forms of social contract, so there are also major risks involved. For Zelikow, the most important thing is that the masses take the reading of those events, the public myth, as true and that the middle class of technical and administrative framework (the 'relevant political community,' its cadre) actively propagates it. In the current crisis, that is actually what we are witnessing: politicians of every color, the media and the columnists, the people who sit on the TV talk shows, they all tell the same story. The majority of the population believes it too. Deceit has thus been upgraded to a priority task for all governments, supported by the new power bloc of intelligence agencies, IT corporations and media.

TRANSNATIONAL NETWORKS

The propagation of a worst-case scenario, like the development of a concept of control, relies on extensive networks of communication which have largely (but not entirely) supplanted the role of the Church, Freemasonry, etc. in the process of ruling class formation. These are, first of all, the networks formed by interlocking directorates connecting large corporations. Peter Sutherland, mentioned in the previous chapter, was an example: he held top posts in Goldman Sachs, BP, and a host of other companies, besides being UN migration chief. Next, confidential, members-only networks such as the Bilderberg conferences and the Atlantic Council, or the Trilateral Commission, the G30 (the group of 30 ex-presidents of central banks), and the World Economic Forum, allow the owners and top executives of these corporations to consult confidentially with key figures from the worlds of politics and the media. These consultative, policy planning networks are in turn interlocked with the networks of corporate directorates.[62]

The planning groups-cum-think-tanks are an important inter-mediary between the actual ruling class and governments. Barack Obama had the backing of the Trilateral Commission and the G30 for his first election as president; eleven members of the TC were appointed to his government, whilst Zbigniew Brzezinski, its co-founder, was its chief foreign policy adviser. The emphasis on a 'Russian threat' and the ploy to seek to remove Putin, had been developed in the TC before being taken up by the Obama team, culminating in the coup in Kiev in February 2014.

If we look at the composition of the directive bodies of these networks, there is no clear pattern conforming to the intel-ligence-IT-media triangle to be found in the Executive Group of the TC; it is a cross-section of the capitalist class. The Steering Committee of the Bilderberg Conferences, on the other hand, does reflect the new power bloc. In addition to Eric Schmidt, for-merly of Google, important new IT companies such as Palantir and IT entrepreneurs such as Peter Thiel sit on it, as do directors of Airbus and various media figures such as the Belgian banker and press magnate, Thomas Leysen, along with representatives of financial institutions such as Lazard, AXA, Deutsche Bank and Investor (the Wallenberg group of Sweden).[63] Henry Kravis, one of the partners of Kohlnerg-Kravis-Roberts (KKR), one of the pioneer hedge funds, is a key figure in Bilderberg and the Trilateral Commission; he is Emmanuel Macron's key patron in the capitalist class.[64]

The Atlantic and global policy networks are also interlocked with regional European ones, like the Belgian-French elite soci-ety, the Cercle de Lorraine. The aforementioned Thomas Leysen is a speaker at its meetings, addressing among others, the Belgian press magnate with whom he owns practically the entire Belgian and Dutch newspaper market, Christiaan Van Thillo, who in turn is also a director of the German publishing group Bertelsmann, one of the six largest media empires mentioned already. Another Belgian member of the Cercle, Count Maurice Lippens, is also a prominent member of Friends of Europe (of which former Belgian prime minister Guy Verhofstadt and former European Commisioner Neelie Kroes are members as well), and so on and so forth.[65]

How these networks also serve to delineate the ruling class consensus is exemplified by the fact that the only mainstream weekly in the Netherlands that publishes critical information on the Covid crisis, is also one of the few not owned by either the Leysen or Van Thillo press conglomerates. In France, *France-Soir* first had to close down as a print publication before it would depart from the otherwise solid media consensus regarding the 'pandemic.' Other French media are mostly owned by the oligarchy. The richest dynasty, Bettencourt (L'Oréal) participates in *L'Opinion*, jointly with luxury goods conglomerate LVMH of Bernard Arnault (no. 2), which also owns the *Le Parisien* and *Les Echos* media groups. *Le Figaro* is owned by aerospace giant, Dassault (fifth richest family); F. Pinault (no. 7), owns *Le Point*; Patrick Drahi (no. 9), *L'Express*, *Libération*, and indirectly, BFM-TV; Xavier Niel (no. 11), the *Le Monde* group (*Courrier International*, *Le Monde Diplomatique*), and so on.[66] The concentration of the U.S. media market following the Telecommunications Act was already mentioned; in Britain, three publishers—DMG Media of Lord Rothermere (the *Daily Mail*, and related titles); Murdoch's News UK (*The Sun*, *The Times*...); and Reach Plc (the *Mirror*, *Express* and *Star* titles)—accounted for 90 percent of print sales, the remainder being shared by the *Guardian*, *Financial Times*, and the *Telegraph*.[67] In Germany, on the other hand, the newspaper market is much less concentrated, which perhaps explains the beginning press criticism of the pandemic in that country.

The oligarchy on both sides of the Atlantic owns media for political reasons, insofar as print publications are mostly net losses. Niel bought *Le Monde* for the amount his fortune fluctuated in one day. Bezos acquired the *Washington Post* from the venerable Graham dynasty to add a voice to his Amazon empire. The input is often coming from the state, including leaks, or from think tanks publishing reports.

The think tanks in turn are heavily subsidized by the military-industrial complex—nowhere more so than in the United States. The top 50 American think tanks received more than a billion dollars from the government and weapons manufacturers in 2019: from the Pentagon (381.7 million), the Air Force, 287.4 million, the Army, 246.3 million, the Department of Homeland Security, 111.2 million and at a distance, from the State Department with 9 million. The main private donors were Northrop Grumman, Raytheon, Boeing, Lockheed Martin, and Airbus—the core of the Atlantic arms industry. This is not surprising since the main beneficiary by far is the dedicated arms and military strategy research institution, the RAND Corporation, which received the bulk of the money (a little over a billion). The others are much smaller but equally bellicose: the Center for a New American Security (with CEO Victoria Nuland, who directed the 2014 coup in Kiev and under Biden, returned to the State Department as Under Secretary) received $8.9 million; the Atlantic Council, with Richard Edelman and the chair of the Munich Security Conference, Wolfgang Ischinger, on its Executive Committee, $8.6 million. The New America Foundation, headed by Anne-Marie Slaughter, a former State Department planning director under Hillary Clinton, is close to the major IT companies, especially Google (Eric Schmidt) and the Gates Foundation. It received 7.2 million. The German Marshall Fund of the U.S., 6.5; Georgetown University's CSIS, where the Biden government recruited Kathleen Hicks as Under Secretary of Defense, 5 million; the Council on Foreign Relations, 2.6; the Brookings Institution, 2.4; the Heritage Foundation, 1.3 and the Stimson Center, 1.3 million.[68]

The intelligence-IT-media bloc outlined above and the transnational networks by which it passes on its power to a broader coalition, thus have joined forces to propagate a new worst-case scenario in response to the threat of an impending 1848. Since in the preceding period the role of a class vanguard in this sense resided with the financial sector, and especially speculative money-capital, the driving ideology of a 'chasing yield,' risk society (compounded by the fear of terrorism), had to be replaced by another. My contention is that it is here from whence the myth of the 'pandemic' derives, serving as the new worst- case scenario in the overall politics of fear, now that 9/11 and 'terrorism' have begun losing their effectiveness. Still according to Zelikow's logic, a formative event was needed to shock people into compliance. That could not be another financial crash, because such an event might pull the entire constellation of Western power into the abyss with it.

CONSOLIDATION OF THE FINANCIAL SECTOR

When a new financial collapse indeed threatened to occur in September 2019, one even surpassing that of 2008, the need to discipline the fraction of money-dealing capital, as Marx calls it, became acute. One of the determining factors leading to the enunciation of the Covid state of emergency, therefore, was to prevent a new financial disaster, because the population was already in a state of tension and possible revolt. There is one further factor that played a role in the *timing* of the Covid crisis and that was Trump's more than likely re-election in November 2020, which I will discuss later. Here we concentrate on the need to consolidate the financial sector and make it more crisis-proof.

'2008' was not just a regular stock market crash or recession. Wolfgang Streeck sees it as the moment when the pyramid of financial remedies by which successive attempts had been made to maintain social peace in the West after the turbulent 1960s and 1970s, finally collapsed and that going forward, democracy would be on hold and government would have to rely on other means.[69] The question, then, was what this would mean for the capital fraction that heretofore had been at the forefront of the power bloc since the 1990s and had even succeeded in directing an apparent recovery—until 2020.

After the collapse of the Soviet bloc and the USSR, the disciplining effect on the public of the Cold War lost its effectiveness and as noted already, speculative capital had come to dominate the financial sector. Profits increasingly depended on exploiting price differences, if necessary by creating them through bubbles, e.g. in real estate. The man who first developed this type of money-dealing at Salomon Brothers, John Meriwether, in 1994 set up, along with two so-called Nobel Prize laureates, a hedge fund of their own, Long-Term Capital Management, LTCM. It promptly went bankrupt in 1998.[70] The regulated banks, meanwhile, ensured that under Bill Clinton's presidency, the Glass Steagall Act, the New Deal legislation that forced depository banks to let go of their investment arms, was reversed so that they could also join the hunt for commercial profit.[71]

In the 1980s an insurance system had been created in which the U.S. government underwrote the risk and the speculator within certain limits was covered. It was followed by an informal insurance system in the 1990s. So when LTCM went bankrupt, the Federal Reserve under Alan Greenspan (a former corporate director of JP Morgan) put together a rescue consortium to contain the effects across the sector. Such bailouts turned each subsequent financial crisis into a prelude to new bubble (in stocks, real estate, etc.), all along continuing the process of the strip-mining of entire economies, to use Jack Rasmus's phrase. Price signals guided investment decisions on the basis of artificial intelligence, with the help of algorithms. Thus, crises occurred, one after another: the Asian crisis of 1998, bankruptcies like those of Mexico, Russia and Argentina, and the dot.com bubble, before the sector finally exploded in 2008.[72]

The story of (re-) insurance practices in which uncertain investments were packaged in derivatives which in turn served as collateral for new credit operations, does not need to be told again here. Increasingly these 'instruments' were traded internationally as securities, often with a component of debt that could never be redeemed, such as the notorious subprime mortgages, granted to people who would not be able to meet their obligations after the initial grace period. Their creditworthiness was invariably underwritten by the credit rating agencies such as Moody's and Standards & Poor—until the subprime mortgages pulled down the house of cards when it came to paying. Initially, the money traders succeeded in getting redress, e.g., through the $700 billion bailout launched by George W. Bush's Treasury Secretary, former Goldman Sachs CEO Hank Paulson, which was paid out in full to banks that had taken (too) great risks.[73]

Nevertheless, with the benefit of hindsight, we can see that the transition to a new financial order began with Barack Obama taking office. The top banks and shadow banks and even big speculators like George Soros and Paul Tudor Jones, had supported Obama's candidacy for president because they feared a revolutionary situation similar to that on the eve of the New Deal in the 1930s. The threat of insurgency, however, faded quickly; by 2010 governments had succeeded in convincing public opinion that not the speculators, but the *bon vivants* in southern Europe had caused the crisis.[74] Greece was then sacrificed to bail out the banks that had speculated with Greek government securities. Timothy Geithner, Paulson's successor in the Obama administration, likewise seemed inclined to bail out the money-dealers. Geithner hailed from the New York Federal Reserve and was close to Citibank. In 2010, two years after the collapse, total assets of the shadow banks were again 20 percent higher than the regulated banking sector. Three years later, shadow banks worldwide again controlled $75.2 trillion (from $26 trillion in 2002). According to Jack Rasmus, one-third of the world's total was registered in the U.S. ($25 trillion), compared to $10.5 trillion in assets in regulated U.S. banks (the 38 largest).[75]

However, the sector had become even more vulnerable because the central banks (the Federal Reserve in the U.S., the Bank of England, and the European Central Bank, ECB), in addition to providing direct support, bought up questionable securities through Quantitative Easing (QE), by which the financial asset investors were again provided with funds for a new round of speculative operations. The public was told that these were enabling investments in the real economy, whereas in reality QE only drove up stock prices, and tempted asset investors into taking even greater risks. Between 2008 and 2015, the U.S. central bank, the Federal Reserve, conducted three rounds of QE, buying up trillions' worth of government bonds and debt securities, in exchange for new capital for financial investors. At the end of 2015, the U.S. central bank's balance sheet was inflated to $4.5 trillion. Similar operations were conducted by the ECB under Mario Draghi (ex-Goldman Sachs) and the Bank of England.[76]

Yet the U.S. central bank is remaining in the lead, with the new Biden government in the process of breaking all records. In total, the U.S. has 'printed' $8 trillion, including the latest 1.9 trillion proposed by Biden. Such an extent of public debt issuance has never before happened. The same is going on in Europe. Meanwhile it is an illusion to think that U.S. inflation has been kept under control at 1.4 percent and interest rates are zero; it is only because economies have been frozen by lockdowns that these figures remain so low. In reality the figure should be around 10 percent. Because the value of the dollar is linked to financial assets, they will go down together, so a different approach is urgently needed.[77]

NEW PROMINENCE OF THE PASSIVE INDEX FUNDS

The structural change that began after 2008, at first surreptitiously, came in the form of a consolidation of the financial sector, with the predatory hedge funds losing ground to a new type of financial investment firm, the passive index fund. These funds are called 'passive' because they are no longer chasing yield left and right but rather invest in large, established companies across the entire economy. Between 2008 and 2019, the passive index funds succeeded in increasing their assets by an additional $4 trillion while actively operating investors, the speculators of the preceding period, in near-perfect mirror image saw their portfolios shrink by more than three trillion. In retrospect, this was the most important mutation; other measures to contain speculative capital were unsuccessful. The Volcker Rule (named after Paul Volcker, the man behind the 1979 interest rate hike that triggered the global debt crisis), was intended to protect banks' equity and deposits, but it was not implemented. It only cost Obama his support on Wall Street.

The advance of the passive index funds initially seemed part of the bail-out, nothing more. When insurance group AIG, which had also entered the trade in packaged securities, ran into difficulties, Secretary Geithner ordered Larry Fink of BlackRock to resolve this crisis, so that large institutions such as Goldman Sachs and Deutsche Bank would not get into trouble. The most striking result was that all money-dealers and especially their shareholders were rescued (only a handful, most notably Lehman Brothers, were sacrificed). In the process, the three major passive index funds, BlackRock, State Street and Vanguard, took the lead (80–90 percent of the market) to the extent that they can no longer be seriously threatened by European or American competitors.[78]

BlackRock got through the crisis fairly unscathed thanks to generous fees for services rendered. It was the record purchase of Barclays Global Investors of the eponymous British bank in 2009 that catapulted it into the top bracket.[79] In the U.S., BlackRock is the majority shareholder of 4 of the 5 largest banks. In the EU, it is the majority shareholder of Deutsche Bank, Dutch ING, the UK's HSBC and Banco Bilbao, and the second largest of BNP Paribas, Unicredit and Banco Sanpaolo. Here it should be remembered that BlackRock itself is also an investment object for the other passive index funds, Vanguard and State Street, as well as for JPMorganChase and other American banks, and Japanese Mizuho. BlackRock is also active in the pharmaceutical industry as a majority shareholder of Pfizer. In September 2016 it organized the takeover of Monsanto, the notorious producer of genetically modified seeds (and in which BlackRock is the third largest shareholder) by Bayer (main shareholder, BlackRock).[80]

This qualitative change of the financial system from the speculative hedge fund model to the pattern of the passive index funds has to some extent restored a long-term perspective in the sector. The passive index funds are not aimed at specific companies but invest across the economy as a whole and thus have an interest in its flourishing in the sense of a high average profit rate.[81] According to the latest data, BlackRock has the largest investment portfolio with $5.4 trillion in capital invested, followed by Vanguard with 4.4, JPMorgan Chase with 3.8, and Allianz with 3.3 trillion. State Street is at no. 8 with 2.4 trillion.[82]

The three passive index funds, as well as the banks and shadow banks that have also moved to broad investment portfolios, are all deeply involved in the IT monopolies — Microsoft, Apple, Facebook, Google and Amazon. But that does not make them a component of the new power block behind the pandemic, because here their 'passivity' (which of course is always relative and possibly temporary—Fichtner and Heemskerk see more active involvement in company management on the horizon) rules out that they take the lead. They are fellow travelers, *not* the organizers of the new class bloc. In this respect they differ from the banks that constituted finance capital around the turn of the 19th to 20th centuries in Germany and Austria, as defined by Rudolf Hilferding. These were combinations consisting of an investment bank with a group of interlocked companies in which the banks called the tune, a system that has only been undone in our time.[83]

Like that model, however, the passive index funds also represent a step forward in the socialization *(Vergesellschaftung)* of the economy that is the prerequisite for its being taken into public ownership *(Sozialisierung)*. Again, the step from actual growth in capital to active action as a class or fraction accompanied the shift. On the eve of the 2016 presidential election, thirteen fund managers and CEOs, including Fink, Jamie Dimon of JPMorganChase, and Warren Buffett of Berkshire Hathaway, signed a statement in favor of a longer-term perspective in financial markets.[84] The aforementioned Group of 30 (G30), which was set up in 1978 with the support of the Rockefeller Foundation and consists of (former) chairmen of central banks, directors of the Bank for International Settlements in Basel, and a few more financial policymakers, share this perspective. Their advice is mainly in the field of regulating the financial sector, most recently with the help of the Systemic Risk Council, established in 2012 to prevent a new 2008.[85]

In 2019, however, another 2008 seemed to be just around the corner. As we can now see, it was only the Covid state of emergency and the lockdowns with their deflationary effect that prevented another crash from happening—for the time being.

A NEW 2008?

The possibility that the financial sector could indeed bring the capitalist economy to the brink of collapse, just like in 2008, nonetheless still came a step closer when the U.S. 'repo' market (from repurchase) threatened to break down in mid-September 2019. This market ensures that there is sufficient liquidity to keep the payment system running. Banks, investment banks and mutual funds acquire this cash with securities as collateral, often U.S. Treasury bonds. The cash is usually repaid within a day. In the fall of 2019, however, an unexpected hike in short-term interest rates threatened to paralyze the repo market, and the New York Federal Reserve stepped in by making dollars available. By March 2020, the Fed had already created $9 trillion in new money to keep the repo market going. However, which banks benefited from this remained a secret.[86]

At the first sign of imminent disruption, however, something else happened: the obviously well-informed executives of large companies decided to leave their posts and, in many cases, sell their shares. In total, 1,332 CEOs had left their posts by October 2019. And whilst it is not uncommon for such an exodus to take place in a recession, to see this happening while big profits are being made and the stock market had reached an all-time high was unusual. The most likely explanation, then, is that the top managers had information that bad weather was coming.[87]

The exodus of CEOs not unexpectedly continued as the Covid scenario came on stream in January. The restructuring of production chains away from China is making the already fragile, disintegrating world economy even more vulnerable, but that is not the chief concern of CEOs, which is self-interest. In the current economy, the bosses of non-financial companies are also primarily investors. The stock market price has to go up and they achieve this by having their companies buy up their own shares, which are usually also part of their remuneration. More than 200 top executives left their posts in the first month of 2020. Finally, in March, a collapse of the stock market occurred, the like of which had not been seen since Black Monday in 1987, vindicating all those who had stepped out in time.[88]

The Covid crisis was again responded to by Quantitative Easing. By the end of May 2020, the G20 states had already collectively spent $7 trillion for tax relief and direct aid. That amounted to more than 10 percent of their joint national product, well exceeding what had been thrown in during the financial collapse of 2008. The deficits that resulted fall in the category of the sort of debt incurred, for example, by Great Britain in the Second World War and will take several generations to pay back. Austerity such as that after 2008 will not be enough.[89] As mentioned, uncontrollable inflation would already have been created by the repeated QE operations, had this not been prevented by locking down the economy.[90]

In the process, the eradication of the old middle class is not an unforeseen side effect of the lockdowns. Under the aforementioned 'screen New Deal' physical shopping is meant to be replaced by online purchases at Amazon and its equivalents, as part of loosening the social fabric by having all transactions and contacts take place digitally. In a 'smart city' every purchase is registered, and the social category of shopkeepers is likely to be wiped out, its market share transferred to Amazon and its kin.[91]

With the combination of QE, secrecy, and increasing stagnation of production, the circulation of capital came to a standstill: the world's 2,000 largest non-financial companies have accumulated financial reserves reaching $14.2 trillion in 2020, against $6.6 trillion in 2010. However, as William Robinson points out, capital cannot stand still indefinitely without ceasing to be capital.[92] At the same time it is not impossible that the power of the intelligence-IT-media bloc is reaching even deeper into the operation of the financial sector. As we shall see, the involvement of MasterCard in various plans of Bill Gates may be an indication that money as we know it may disappear. The company is closely associated with the IT giants and in 2018 it was revealed that it had sold (for an unspecified amount) credit card data to Google. MasterCard has also partnered with Visa, Citibank, the Gates Foundation, USAID, and other companies and institutions to fund the Better Than Cash Alliance for the abolition of physical money. In February 2021, MasterCard announced that it will also handle crypto transactions through their network in the course of that year.[93]

NATURE AS AN EXPLANATORY FACTOR FOR THE DEPRESSION

While stock prices rose again after mid-March 2020, economic and monetary policy were leaving all economic logic behind. Those who had benefited from the rescue operations remained anonymous; the banks were not to be blamed. The 'measures,' i.e. the lockdowns in the name of the pandemic were well on the way to turning a financial denouement into a global depression.[94] Only through such a depression can a renovated capitalism be launched, as happened in the 1930s. But at that time, the mobilization of democracy and the confrontation with fascism played a defining role, whereas in the current crisis, even nominal, formal democracy and basic constitutional rights have been put on hold. Whether this will again result in war is a terrifying uncertainty.

Meanwhile, the pandemic is significant in another sense, too, as the unparalleled social and economic havoc can be attributed to 'nature.' This time, there is no need to fool the public by blaming the Greeks for the crisis, although there is a lot of complaining going on about Italy. For now, we have the virus to explain it all. As early as the 19th century, when the counterattack was launched against the workers' movement and the classical labor theory of value, which under the pen of Marx had been given a revolutionary twist, the ideologues of the capitalist class came up with the idea of making nature the explanatory factor. In addition to the idea of the declining fertility of arable land, on which the new, subjective value theory of marginalism was based, Stanley Jevons (one of its proponents) also came up with the idea that the business cycle was based on sunspots, so nothing could be done about that either.[95]

Yet as indicated above, the concrete exercise of power, in capitalism as in other types of society, is always the result of the relations of forces between social classes and within them, class fractions. In 2020, the year in which the Covid crisis was declared, there was another major bone of contention in that regard, and that was the impending reelection of the populist, Donald Trump, in the United States. Whatever Trump represents, it is not the triangle of the intelligence world, the IT corporations and the media, nor did he gain much credence in the transnational and U.S. networks within which they have entrenched themselves and built their alliances.

NATIONAL POPULISM: SHOCK TROOP OR OBSTACLE?

If the timing of the Covid state of emergency has not been occasioned by the impending financial collapse, or at least, not solely, then the rise of national populism should be taken into consideration as well. More particularly, the fact that populist leaders had been elected to the highest office in the two most important countries of the Western Hemisphere, the U.S. and Brazil, cannot be overlooked in this connection. This again reverts back to what I see as the main driver of the Covid state of emergency, the threat of a new 1848.

According to the classical definition, based on the experience in Latin America (Perón in Argentina and some of his contemporaries), populism denotes that 'the people' are constantly invoked as a source of resistance against an unspecified 'elite,' but the people themselves are not supposed to actually take action or become an active force in any shape or form.

Nationalism has always been an important component of populism. This is associated with its denial of class differences, and is expressed, for example, in corporatism (professional and industry-oriented socio-economic organization). In addition, national populism perceives society as an *organism* that may or may not be 'healthy' and from which foreign elements must be removed. The claim is that the nation's original health— whether its national or its racial purity, or in a more disguised form, 'our culture' or 'our values'—has been affected by alien ideas or actual immigration. This organic conception of a quasi-natural 'people' obtained its extreme translation in fascism and is the logical counterpart of the theory of class struggle of Marxism. Hence the paradox that although Marxism has practically disappeared from intellectual life in the West, populists never tire from fulminating against it, as against anything else associated with the historic left. Never mind that the actual political left has long been absorbed into the broad center of politics as the 'center-left,' and gladly lends a helping hand in executing the policy choices of transnational capital and the internationalized state, as in the Covid state of emergency today.

As a result, the economic interests of the mass of the population are no longer represented politically, leaving national populism, with its campaigns against immigration (rather than against war or poverty), against Islam, against 'Europe,' to present itself as the resistant alternative. In a few cases, national populists such as in France (Marine Le Pen) have adopted the former Keynesian program of the left, viz., public investment, the monitoring of purchasing power, and active anti-cyclical policies, but this is an exception. Mostly they act as shock troops for the neoliberal project by destroying the social fabric and preventing solidarity between the majority déclassé population and the immigrant or minority sub-proletariat. The same applies to their attitude towards cheap labor, whether from Eastern Europe in the EU, or from Mexico and Central America in the U.S.

National populism is therefore much less remote from the broad political center than is often assumed. Within the ruling class there is even less of a distinction, as there is no real fault-line with the IT oligarchs backing the center; the most one can say is that the backers of populism are marginal compared to the core intelligence-IT-media bloc. Facebook is a data provider to the IT entrepreneurs who, using artificial intelligence and Big Data, were a factor in the election of Donald Trump and the Brexit referendum.[96]

A key player in this field is Robert Mercer, a billionaire who made his money with his hedge fund, Renaissance Technologies, which specializes in automated arbitrage and asset trading. Mercer has spent tens of millions on political causes, including a Heartland Institute he set up to combat the idea of climate change. This, too, is a central theme in the propaganda of the populists, enabled by the false environmental rhetoric of the broad center, which primarily seeks to keep capitalism out of the discussion. Mercer supported the Brexit referendum campaign of his friend Nigel Farage and the campaign of Trump, and is one of the owners of the national populist *Breitbart News Network,* named after Andrew Breitbart, who died in 2012. Breitbart, which had been launched by its founder with the mission to 'reclaim our culture,' has offices in the U.S., Jerusalem and London. On Mercer's recommendation, editor-in-chief Steve Bannon became Trump's campaign manager and (briefly) White House strategist.[97]

Subsequently, Bannon turned up in Europe, where he founded a small organization in Brussels, The Movement, in September 2018. His right-hand man was the chairman of the Belgian Parti Populaire, Michael Modrikamen. The aim was to build a victorious coalition of the far right in the 2019 European elections and beyond. To this end, Bannon wanted to apply proven American-style methods such as polls, data analysis and intensive social media campaigns. Bannon's team had an initial meeting in Rome with Italian Interior Minister Mateo Salvini, leader of the populist Lega, who eventually was expelled from the coalition government with the 5-Star Movement. The duo then spoke with Hungarian Prime Minister Viktor Orbán.[98]

The technical side of electoral manipulation was entrusted to Strategic Communications Laboratory (SCL) and its subsidiary, Cambridge Analytica, of which Steve Bannon is vice president. In the propaganda campaign that accompanied the foreign military interventions in the War on Terror, these techniques were further perfected. For its election operations, SCL worked with the U.S. State Department's Global Engagement Center (GEC) on 'audience analysis' through Facebook ads targeting certain age groups in a particular country and collecting *'likes'* to draw up personality profiles.[99]

We see here that there is no essential incompatibility between mainstream politics and the media, and these networks. SCL works with the Dutch military intelligence center against the Covid opposition, and the head of SCL Elections is Mark Turnbull, a former advertising executive of Saatchi & Saatchi who worked for Margaret Thatcher, among others. Turnbull also made the videos with a fake Bin Laden and is working closely with the U.S. and UK defense ministries to launch anti-Russian campaigns. According to Nafeez Ahmed, all this is no longer straightforward PR, but the expression of the increasing radicalization of the Anglo-American military.[100]

Bannon also assisted with the election campaign in Brazil of the national populist, Bolsonaro. In August 2018, one of Bolsonaro's sons, Eduardo, met with Bannon in New York to discuss this commitment. Taking advantage of Cambridge Analytica's experience in the 2016 U.S. election, Bolsonaro's campaign targeted undecided voters in the northeastern states, as well as women voters, with a barrage of fake news about the PT, the Workers' Party of Lula and Rousseff, which had adopted many if not all of the features of the hated broad center. 'Theoretically' this was supported by a campaign against 'Cultural Marxism,' which refers to the protection of migrants and ethnic minorities and which is also in the spotlight of populists with intellectual pretensions in Europe. Incidentally, Bolsonaro withdrew from public exposure and debate following a knife attack on him in September 2018.[101]

Meanwhile his campaign spread fake news through Whatsapp groups, such as that under a PT government, the state would henceforth decide on the sex of babies, something that according to surveys was in fact believed by 70 to 80 percent of the target group. In addition, Bolsonaro's election was due in large part (like Trump's in the U.S.) to the support of the evangelical churches, which preach conservative social values through their influential TV channels. About a quarter of the Brazilian population is associated with these evangelical movements, which incessantly rage against the 'left,' meaning the privileged urban cadre and their extremely liberal views—although unlike most of Europe, in Latin America there remains a progressive left rooted in the labor movement, as well as Marxist influence among intellectuals. The Bolsonaro campaign meanwhile successfully attacked sex education in schools, as well as gender studies and feminism.[102] He took office as president on 1 January 2019.

National populism, then, is not so much the autonomous expression of an angry population that has taken to the streets on its own account, but a slick PR machine that exploits this anger with the help of the most advanced IT methods to influence behavior. This machine overlaps, via Facebook for example, with operations of the same type but then for the benefit of the broad center, behind which hides a much more powerful bloc of ruling class forces led by the big IT monopolies and the mainstream media, the mainstream of the Atlantic military-industrial complex and the intelligence world, as well as the financial power of Wall Street and the City.

Britain's departure from the EU, Brexit, likewise fed on popular anger, which the Labour Party led Jeremy Corbyn was not able to assimilate and translate into a coherent policy, as it was itself divided on the issue. According to Guilluy, there was a lot more going on here than just the question of whether to say goodbye to Europe or not; the Brexit vote was a signal that reminded the political class of the existence of an 'indigenous' population made redundant by the internationalization of capital and labor.[103] The Tories on the other hand, like the Republicans in the U.S., were certainly divided too but successfully adapted to the populist tide without entirely surrendering to it, thus escaping the crisis of the center-left.

In addition to the overlap between the center and national populism in IT and election technology, another area where they differ only in nuances but share the same basic attitude, is the assessment of Israel and the ongoing colonization of the occupied Palestinian Arab territories. The populists typically adopt a position of unbridled support for this policy. While the governments of the broad center still show some restraint, the populist leaders are in favor of the annexation of the remaining West Bank. Whether Bolsonaro's love affair with the Likud administration was even more intense than Trump's, who broke half a century of Western politics by relocating the U.S. Embassy in Israel to Jerusalem, encouraging new settlements, expropriations, and ethnic cleansing, is difficult to determine.

One consequence of the Trump and Bolsonaro victories is that what remains of progressive forces in the U.S. (and the rest of the West) and Brazil, tends to judge the pandemic and the lockdowns, as well as the available medications, through the prism of what these populist demagogues say and do about it. With the Biden presidency it would seem that the mask-wearing forces have regained the initiative, for as Patrick Zylberman already indicated, the public's attitude to sanitary measures comes about through the matrix of political loyalty versus opposition.[104] This is illustrated by the response to the anti-malarial drug, hydroxychloroquine (HCQ), which in combination with zinc and an antibiotic has proven to be a cheap and effective medication against Covid-19 but has been fought tooth and nail by the pharmaceutical industry because it would make all kinds of experiments with genetically modifying the human immune system superfluous and prevent the licensing of intravenous gene therapies. Because Trump and Bolsonaro spoke out in favor of HCQ, however, an 'anti-fascist' consensus was formed claiming that the drug was meaningless or even dangerous, including in circles of highly qualified specialists, who thus expressed their profound dislike of these leaders.

Indeed, the campaign to prevent Trump's re-election in 2020 was fueled by attacks on his statements about HCQ, and perhaps the timing of the measures in response to the corona outbreak, or possibly even the outbreak itself, were motivated by that goal.

THE COUP AGAINST TRUMP

The central thesis of this study is that the Covid crisis provided an avenue of response to the growing unrest among the world's population seized upon by the intelligence-IT-media bloc that enabled it to implement an authoritarian state equipped with the full array of new digital capabilities for permanent surveillance. Here I look at the issue of whether the impetus for unleashing the long-planned pandemic scare was the looming financial crash or to make sure that Trump's reign would end in disarray, or even remove him by force if need be. But whilst, as we will see, the Covid crisis simulacrum had been worked out long before, few had foreseen that someone like Trump would come to occupy the White House.

Trump's election as president is the first example in the U.S. of an outsider who captured the highest office after bypassing the power bloc, benefiting from universal suffrage and his own money. Normally, candidates who do not have a prior mandate from the dominant fractions in the ruling class are eliminated at some point and this is well-documented for the United States.[105] In the Netherlands a particularly dramatic case occurred when it appeared that the national populist, Pim Fortuyn, was heading straight for the prime minister's office on a platform that included a plan to reorganize the Dutch armed forces and cancel the order for the F-35 jet fighter. He was assassinated in 2002. His spokesman, a minor military-industrial consultant who nonetheless attended a Bilderberg conference, replaced Fortuyn at the head of his parliamentary group, which went on to vote for the plane before dissolving again.[106] Subsequent Dutch populists, such as the anti-Islam crusader, Geert Wilders, whose breakaway from the mainstream liberal party was supported by the Israeli embassy, or Thierry Baudet, who has the backing of Protestant fundamentalist real estate interests, have never equaled Fortuyn's charisma and meteoric rise.[107]

In the United States, the aversion to 'Washington' has often been exploited during election time, but presidents so elected have always fallen in line after taking office. Trump, on the other hand, was a real outsider in the White House. According to polls, the moderate socialist, Bernie Sanders, who advocated taking on Wall Street, would have defeated Trump—yet another proof that the rejection of mainstream politics would not necessarily mean a right-wing nationalist choice—if such a candidate were indeed able to secure the party nomination. However, the Democratic Party apparatus undermined Sanders' campaign (as it would do again in the 2020 campaign against Biden) in the primaries to put Hillary Clinton, the candidate of the power bloc, against Trump, only to be beaten against all expectations.

His outsider position nonetheless denied Trump sufficient control over the U.S. state apparatus, especially in the foreign policy domain. Not a single figure on his foreign policy team (John Bolton, Mike Pompeo, Nikki Haley, James Mattis) emerged from the aforementioned networks such as Bilderberg and the Trilateral Commission, the Atlantic Council, or U.S. elite networks such as the Aspen Institute, the Council on Foreign Relations, or the Brookings Institution. The three previous presidents, both Republican and Democrat, had always tapped into these reservoirs for their cabinets, as did Biden upon his election.[108]

As a multi-millionaire who immediately introduced tax cuts for the very rich, Trump could count on some goodwill from Wall Street (initially he even placed three Goldman Sachs bankers in key positions), but in that respect too, Trump is an outsider: the financial base of his real estate empire is said to be the Russian mafia.[109] But the attempts to dislodge Trump did not use that connection, and instead based their impeachment strategy on the unlikely accusation that he was on Putin's leash though Putin, according to the U.S. embassy in Moscow, probably has little or no control of the elements in the security services entwined with the mafia.[110]

With his base in the ruling class already weak, Trump's blue-ribbon advisers from the business world, soon walked out again. The Strategic and Policy Forum (Fink, Dimon of JPMorganChase, the CEOs of WalMart, Boeing, IBM, and a few others) distanced themselves from the president after the Charlottesville race riots in August 2017. Two months earlier, Bob Iger of Disney and Elon Musk of Tesla and SpaceX had already stepped down in protest over the U.S. withdrawal from the Paris climate agreement.[111]

Trump did not start new wars, not necessarily out of inherent peacefulness but because he wanted to repair the dilapidated infrastructure at home and repatriate economic activity from overseas. Although he partially succeeded, he was too unpredictable for the ascendant power bloc of intelligence agencies, IT monopolies, and media to allow him to spend another term in the White House. The fact that the Covid-19 pandemic was announced in the U.S. presidential election year cannot be separated from this. The social and economic disruption caused by the contradictory, often state-level measures and the impossible position of a president who did not act according to the Covid script—besides what half the U.S. population reasonably regards as a voting fraud—then helped the hapless Joe Biden to victory. However, steps were also taken to remove Trump by force if necessary.

In an extensive report in *Newsweek* in March 2020, at the height of the Covid crisis and with the Trump re-election on the agenda, a list of measures, which might be taken if the United States would become 'ungovernable,' was discussed in detail.[112] At the time the commander in charge of implementing the Continuity of Government package discussed in the previous chapter, was Air Force General Terrence J. O'Shaughnessy. He was also Commander of the U.S. Northern Command (NORTHCOM), established after 9/11 and including Canadian airspace. The NORTHCOM commander has three scenarios available for a state of emergency, whilst the president and his government are being evacuated to a location in Maryland.

Incidentally, these plans went completely awry on 9/11 because most of the prescribed procedures were either ignored or deliberately pushed aside. The COG Commission subsequently set up to re-examine them received little attention in the U.S. Congress; oddly enough there was little interest in the question of who would exercise authority in a real emergency. Even so, COG measures are practiced every year in the Capital Shield exercise, simulating a nuclear or biological surprise attack, as well as on such occasions as the State of the Union address and presidential inaugurations. The core scenario is always that law enforcement breaks down and the army has to intervene. [113]

All presidents since Eisenhower have renewed the COG provisions with their signature, including Obama; remarkably enough, he did so only in 2016, his last year in the White House. The document in question was Presidential Policy Directive 40 and was classified. But then, in January 2017, just days before Trump was due to take office, Federal Continuity Directive 1 was issued as well. This document, still according to *Newsweek*, specified the circumstances under which authority would pass to the military to ensure the exercise of government functions. The Directive laid down that military force was to be used to quell 'insurgency, domestic violence, unlawful assembly or conspiracy' in the event that civil authorities would be unable or unwilling to deal with the disturbances properly.

It is difficult to determine to what extent an element of a coup d'état was already at issue here. For example, Federal Continuity Directive 1(again, this was issued a few days before Trump was to take over) stipulated that in the event that a U.S. administration would fail to 'demonstrate leadership that is visible to the nation and the world ... and preserve the confidence of the American people,' a substitute authority, that is, a military command, was to take over. Whether the public would have allowed such a takeover of power on the heels of an election or not, the question is: how would the incoming president have interpreted these provisions, which in all likelihood would not have been withheld from him? Was it an intimidation to let him know that the military and the security forces were ready to intervene in the event of any major deviation from the course desired by the ruling power bloc?

Under the terms of the Department of Defense, military commanders have the power to act independently, given the circumstances, if 'the formally appointed local authority is unable to control the situation.' Such conditions include widespread disturbances involving large numbers of victims and extensive damage to property. According to the report in *Newsweek*, the military commander may then institute the state of war, but authority must be returned to civilian hands as soon as possible. These provisions were formalized by the Joint Chiefs of Staff in October 2018, reminding commanders that they were entitled to temporarily establish military rule. As with the Covid crisis itself, the question always arises: why these preparations, especially at this juncture?

In November 2019, when various exercises and simulations were already underway that anticipated the 'pandemic,' an exercise was held in Washington by the Cybereason company, Operation Blackout.[114] The exercise investigated how an election can be hacked by manipulating equipment and falsifying electoral lists. Participants included the U.S. Secret Service (the president's bodyguard), the Department of Homeland Security, the FBI, and the Arlington, Virginia, Police Department, who were then attacked by a group of hackers, security industry professionals, and academics. In another exercise in 2019, Clade-X, (to which I will return in Chapter 6), Tara O'Toole, at the time responsible for science and technology in the Department of Homeland Security, and other members of the Dark Winter team (the June 2001 bio-defense exercise) were involved. Here the scenario was that there would occur a bio-terror attack in the second half of 2020, which would be responded to by introducing a state of emergency according to COG regulations. O'Toole thereafter moved to the CIA's investment arm, In-Q-Tel, where she became executive vice-president.[115]

On the first of February 2020, General O'Shaughnessy, the commander of the Continuity of Government apparatus and NORTHCOM, was ordered by the Pentagon to implement the national pandemic plans (40 days before the WHO announced a pandemic). That was after Secretary of Defense Mark Esper had signed the orders for NORTHCOM and a number of military units on the U.S. East Coast to be ready for deployment in the event that extraordinary operations were needed that could actually involve war on U.S. soil, to assist civil authority, or to secure Washington under COG provisions. A month after Esper, a typical representative of the military-industrial complex (he was a lobbyist for Raytheon from 2010 to '17), had given this order, a series of interventions by high-ranking soldiers, often retired generals, followed. All were clearly intended to discredit the president and undermine his chances for re-election. More specifically, once the Covid crisis dominated all front pages, the letter-writers focused on questioning Trump's authority in light of his statements about the 'pandemic.'[116]

On 9 April 2020, an article appeared on the website The American Mind under the telling headline, 'The Coming Coup,' suggesting that this was how Democrats interpreted the impending election, and specifying that the Internet would be the battlefield. Three weeks later, on 1 May, the *Washington Post,* organ of Amazon owner Jeff Bezos, wrote that a 'Democratic group' was using web technology developed by DARPA to combat ISIS propaganda, to disqualify Trump's statements regarding corona. The 'Democratic group,' DefeatDisinfo.org, was run by none other than former JSOC commander Stanley McChrystal, whose consultancy as we saw was meanwhile playing a lucrative lead role in the cities and states where lockdowns had been instituted. The aim was to use more than a million 'influencers' on social media (an instance of Zelikow's 'relevant political community'), and using artificial intelligence to identify target audiences, to undermine Trump's misguided corona statements by saturating the Internet with the 'right' views. This was also the goal of Event 201, the coronavirus exercise in October 2019, to which I will return at length later and where the hardly new phrase 'the new normal' was launched.

The Black Lives Matter movement following the death of George Floyd under the knee of a white policeman sparked widespread disturbances, including the toppling of statues, looting and arson, disturbances that may well have been embellished (perhaps indicated by their significant white presence) by the anger over the lockdowns in the United States. Biden's candidate for Vice President, California State Attorney Kamala Harris, stated on TV that the BLM movement should continue till election day and if need be, beyond it.

'Color revolutions,' which the U.S. has carried out in so many other countries, from the Philippines to Yugoslavia and Ukraine, require demonstrators out in the streets. Situations where BLM and Trump supporters were facing off against each other became widespread, superseding the vast #MeToo march on Washington that had marked Trump's 2017 inauguration.

When Trump wanted to use the Insurrection Act to deploy the military against the ongoing destruction, Defense Secretary Esper refused to cooperate, and the president came under attack from senior military officers. On 3 June 2020, Esper's predecessor, Marine General James Mattis, launched an attack on Trump and praising the Black Lives Matter movement. Mattis, who resigned in 2018 over plans by Trump to withdraw U.S. troops from Syria (which the president failed to bring about), called BLM a healthy and unifying force and pointed to the lack of mature leadership in the White House. *Foreign Policy* underscored Mattis's statements a day later, on 4 June, in an article entitled 'Generals denounce Trump's protest crackdown plan.' It dismissed the Insurrection Act of 1807 as an antiquated law, two centuries old. In fact, this law had been used 22 times since it was passed, from Jefferson and Woodrow Wilson to Lyndon Johnson and George H.W. Bush, to quell riots and even labor disputes. As Richard H. Black, former chief of the Army Criminal Law Department at the Pentagon, outlined in an interview on *Amazing Polly*, the novelty was that this president was denied the opportunity to use it.[117] In the week following the *Foreign Policy* publication, General Colin Powell, former National Security Adviser and Secretary of State, likewise made a statement that Trump was the first president *not* uniting the country, etc. etc.

The military thus denied Trump the right to use the Insurrection Act against serious disturbances after Esper had de facto abolished Trump's role as commander-in-chief of the armed forces by disobeying the president's order. In an open letter in *DefenseOne* magazine, two former officers then called on the Chairman of the Joint Chiefs of Staff to ensure that if Trump were unwilling to leave the White House on 21 January 2021, he would be forced out by military means. This was not the first suggestion to that effect in the magazine. The scenario that Trump might win on election day but be defeated by postal votes afterwards and would then refuse to leave, was spread on Twitter by the Axios news agency and a Democratic data agency, Hawkfish, linked to Bloomberg. In an MSNBC broadcast, Hillary Clinton impressed on the Democratic campaign that Biden under no circumstances should concede, an idea that originated with the Transition Integrity project, which had run a simulation to see what might happen next in that case.[118] Hillary called on the IT oligarchs, Zuckerberg of Facebook, Reed Hoffman of LinkedIn (with his fact-check machine, Acronym), Bill Gates (who also had entered the fact-checking field himself) and others to steer the opposing information flows on the Internet in the right direction and not to wait for things to go wrong.

On 10 October, *AsiaTimesOnline* gave an overview of the populist strongmen who a year or so before had seemed to be on course to establishing their power permanently. However, Trump, Erdogan, Bolsonaro, Duterte and Salvini (Putin and Xi Jinping were also included in this list) were all on their way out, according to *AsiaTimes*. About Trump, the correspondent in Washington reported that the president had become a victim of the Covid crisis, contending that by appearing in public when he should have been hospitalized, he had even lost the trust of the Secret Service, which as we saw had practiced the possible hacking of the election in Operation Blackout.[119]

Whether or not an electoral fraud that brought the stammering Biden into the White House was indeed organized from the U.S. Embassy in Rome, which with a space satellite of the Italian arms company, Leonardo (of which William Lynn III, former Deputy Secretary of Defense in the Clinton administration, is the CEO), broke into the Dominion voting machines, I will leave aside. However, the documentation from the Italian side deserves to be taken seriously and there is a long tradition of investigating 'deep state' operations in this country.[120] If proved to be the case, it would have been a bitter pill for Trump, who had given Israel an unprecedented free hand, that the Zionist network in which the U.S. Ambassador in Rome, Lewis Eisenberg, is a key figure, helped Biden to victory out of dissatisfaction with his predecessor's insufficient bellicosity. The same would apply to MI6, the intelligence agency of Trump's presumed buddy, Boris Johnson, which was also involved. The fact that NATO and the Israeli military had been alerted by U.S. intelligence about a disease outbreak in Wuhan as early as the second week of November 2019, reminds us that they had had a year to prepare if they really played the roles attributed to them in the Italian scenario. However, according to an Israeli source, their warning concerning the outbreak was ignored by the White House.[121] Trump himself and his entourage preferred to blame 'communist' Chinese and Venezuelan interference as the sources for the startling vote count reversal, and thus may have missed the opportunity to uncover its source.

Ultimately, the planned 'color revolution' was successfully implemented on 6 January. Trump was effectively deposed by means of a mass rally of his supporters who, upon gathering near the Capitol, were invited in by armed guards (videos show how the protesters are being gestured to come in and were then shown the way, walking in an orderly file past the security agents). A media storm promptly erupted over the supposed attack on democracy, amidst strident calls for legislation to prevent any repetition of 'domestic terrorism.' Thus, a further step toward the conversion of the U.S. into an authoritarian state, which in the aftermath of 9/11 had already taken a huge step forward with the Patriot Act, was taken. At the last minute, an impeachment procedure for inciting violence was initiated against Trump to make him ineligible forever, but it failed.[122]

The United States federal system has meanwhile enabled U.S. states to become a beacon of freedom. Since the states retain far-reaching autonomy (which, it will be remembered, was seen by Huntington as a weakness against the ability to quell rebellion), a growing number of them have ended the Covid measures. The states coming out of lockdown such as Texas and Mississippi, and Florida, which never had a lockdown of any significance in the first place, compare favorably in terms of victims with those like California, which introduced one of the toughest lockdowns of all.

This brings us to the question of why and how the scenario of a pandemic was preferred over others as the next installment of the politics of fear.

Endnotes

1. See my *The Making of an Atlantic Ruling Class* (London: Verso, 1984 [reprint with new preface, 2012]).

2. Jeffrey T. Richelson and Desmond Ball, *The Ties That Bind: Intelligence Cooperation between the UKUSA Countries—The United Kingdom, the United States of America, Canada, Australia and New Zealand*, 2nd ed. (Boston: Unwin Hyman, 1990 [1985]), pp. 135–44.

3. James Bridle, *New Dark Age: Technology and the End of the Future* (London: Verso, 2018), pp. 27, 30.

4. Michel Alhadeff-Jones, 'Three Generations of Complexity Theories: Nuances and ambiguities,' in Mark Mason, ed., *Complexity Theory and the Philosophy of Education* [foreword, M. Peters] (Chichester: Wiley, 2008), pp. 65–66. 'Algorithm' comes from the bastardized name of a Persian mathematician.

5. Mariana Mazzucato, *The Entrepreneurial State: Debunking Public vs. Private Sector Myths* (London: Anthem Press, 2014), p. 63; Yasha Levine, *Surveillance Valley: The Secret Military History of the Internet* (New York: Public Affairs, 2018), pp. 7, 61–62, 73.

6. Paul Boccara, *Transformations et crise du capitalisme mondialisé. Quelle alternative?* (Pantin: Le Temps des Cérises, 2008), p. 80; Duccio Bassosi, *Il governo del dollaro. Interdipendenza economica e potere statunitense negli anni di Richard Nixon (1969–1973)* (Florence: Edizioni Polistampa, 2006), pp. 3–4.

7. Nafeez Mossadeq Ahmed, 'How the CIA made Google: Inside the secret network behind mass surveillance, endless war, and Skynet,' *Insurge Intelligence*, 22 January 2015 (online).

8. John Bellamy Foster and Robert W. McChesney, 'Monopoly-Finance Capital, the Military-Industrial Complex, & the Digital Age,' *Monthly Review*, 66 (3), 2014, p. 17.

9. Levine, *Surveillance Valley*, pp. 147, 150; Ahmed, 'How the CIA made Google.'

10. Levine, *Surveillance Valley*, pp. 158, 161; Shoshana Zuboff's bestseller, *The Age of Surveillance Capitalism: The Fight for a Human Future at the New Frontier of Power* (London: Profile Books 2019), deals only with the civilian economic aspect of the Internet.

11. Mazzucato, *The Entrepreneurial State*, pp. 105–7; Ahmed, 'How the CIA made Google.'

12. Levine, *Surveillance Valley*, pp. 175–76, 181.

13. François Chesnais, *Les dettes illégitimes: Quand les banques font main basse sur les politiques publiques* (Paris: Raisons d'agir, 2011), pp. 37–38.

14. Jan P. Nederveen Pieterse, 'Political and economic brinkmanship,' *Review of International Political Economy*, 14 (3), 2007, pp. 467–86.

15. James Jorgensen, *Money Shock: Ten Ways the Financial Marketplace is Transforming Our Lives* (New York: American Management Association, 1986), pp. 95–96; Chen Min; Mao Shiwen, and Liu Yunhao, 'Big Data: A Survey,' *Mobile Network Applications*, 19, 2014, p. 174.

16. Susanne Soederberg, *Debtfare states and the poverty industry: Money, discipline and the surplus population* (London: Routledge, 2014).

17. Jack Rasmus, *Systemic Fragility in the Global Economy* (Atlanta, Georgia: Clarity Press, 2016), pp. 217–22; Anastasia Nesvetailova, *Financial alchemy in crisis: The great liquidity illusion* [with R.P. Palan] (London: Pluto Press, 2010), pp. 9–13.

18. Michel Foucault, *Surveiller et punir. Naissance de la prison* (Paris: Gallimard, 1975), p. 216; Stephen Gill, *Power and Resistance in the New World Order* (Basingstoke: Palgrave Macmillan, 2003), pp. 191–92

19. Nick Dyer-Witheford, *Cyber-Proletariat: Global Labour in the Digital Vortex* (London: Pluto Press and Toronto: Between the Lines, 2015), pp. 95–96.

20. William I. Robinson, 'Global Capitalism Post-Pandemic,' *Race & Class*, 62 (2), 2020, pp. 3–4.

21. Walter Van Rossum, *Meine Pandemie mit Professor Drosten: Vom Tod der Aufklärung unter Laborbedingungen* (Neuenkirchen: Rubikon, 2021), pp. 234–35.

22. Americans for Tax Fairness and Institute for Policy Studies, 'Net Worth of U.S. Billionaires Has Soared by $1 Trillion to Total of $4 Trillion Since Pandemic Began,' *Global Research*, 10 December 2020 (online).

23. Catherine Austin Fitts, 'The Injection Fraud—It's Not a Vaccine,' *Global Research*, 1 January 2021 [orig. 28 May 2020] (online).

24. Joseph A. Camilleri and Jim Falk, *Worlds in Transition: Evolving Governance Across a Stressed Planet* (Cheltenham: Edward Elgar, 2009), pp. 427–29.

25. Citations from Tim Schwab, 'Bill Gates's Charity Paradox,' *The Nation*, 20 March 2020 (online).

26. Fred Moody, *I Sing the Body Electronic: A Year with Microsoft on the Multimedia Frontier* (London: Coronet, 1995), p. 81 & passim; David Klooz, *The Covid-19 Conundrum*, n.p. [Apple Books], 2020, p. 52.

27. Jane Mayer, *Dark Money: The Hidden History of the Billionaires Behind the Rise of the Radical Right* (New York: Doubleday, 2016), pp. 317–18, 320.

28. Glenn Greenwald, 'With new D.C. policy group, Dems continue to rehabilitate and unify with Bush-era neocons,' *The Intercept,* 17 July 2017 (online); *Alliance for Securing Democracy,* 2020 (online); Stephen Kinzer, 'In an astonishing turn, George Soros and Charles Koch team up to end U.S. "forever war" policy,' *Boston Globe,* 30 June 2019 (online).

29. Thomas Piketty, *Capital in the Twenty-First Century* [trans. A. Goldhammer] (Cambridge, Mass.: Harvard University Press, 2014), p. 463.

30. Levine, *Surveillance Valley*, p. 127; Foster and McChesney, 'Monopoly-Finance Capital,' p. 22.

31. Peter Phillips, *Giants: The Global Power Elite* [foreword, W.I. Robinson] (New York: Seven Stories Press, 2018), pp. 264–65, 262–74.

32. Tim Schwab, 'Journalism's Gates Keepers,' *Columbia Journalism Review*, 21 August 2020 (online); Van Rossum, *Meine Pandemie mit Professor Drosten*, p 145.

33. Van Rossum, *Meine Pandemie mit Professor Drosten*, pp. 137, 140–41.

34. Phillips, *Giants: The Global Power Elite*, pp. 284–85, 298.

35. Paul McKeigue *et al.,* 'Briefing Note on the Integrity Initiative,' *Working Group on Syria Propaganda and Media*, 21 December 2018 (online).

36. Mohamed Elmaazi and Max Blumenthal, 'The Integrity Initiative and the UK's Scandalous Information War: The carefully concealed offices of a covert, British government-backed propaganda mill at the center of an international scandal the mainstream media refuses to touch,' *MintPress*, 18 December 2018 (online).

37. See my *Flight MH17, Ukraine and the new Cold War: Prism of Disaster* (Manchester: Manchester University Press, 2018), pp. 135, 139, 143, 155.

38. *McKinsey & Co.*, 'Telehealth: A quarter-trillion-dollar post-COVID-19 reality?,' n.d. (online).

39. Naomi Klein, 'Screen New Deal: Under Cover of Mass Death, Andrew Cuomo Calls in the Billionaires to Build a High-Tech Dystopia,' *The Intercept*, 8 May 2020 (online).

40. Vandana Shiva, 'Bill Gates' Global Agenda' [excerpt from *Oneness vs. the 1 %*], *Children's Health Defense*, August 2020 (online).

41. Flo Osrainik, *Das Corona-Dossier: Unter falscher Flagge gegen Freiheit, Menschenrechte und Demokratie* [foreword Ullrich Mies] (Neuenkirchen: Rubikon, 2021), p. 259.

42. Moderna.com, 'mRNA Platform: Enabling Drug Discovery & Development,' n.d. (online).

43. Austin Fitts, 'The Injection Fraud—It's Not a Vaccine.'

44. Akela Lacy, 'How New York Gov. Andrew Cuomo is Using the Pandemic to Consolidate Power,' *The Intercept*, 28 April 2020 (online); and Anuja Sonalker, CEO of Steer Tech, cited in *The Intercept Newsletter* (online).

45. Gerd Junne, 'Das amerikanische Rüstungsprogramm: Ein Substitut für Industriepolitik,' *Leviathan: Zeitschrift für Sozialwissenschaft*, 13 (1), 1985, pp. 23–37.

46. Robinson, 'Global Capitalism Post-Pandemic,' p. 5.

47. Giorgio Agamben, 'Bioscurity and Politics,' *Quodlibet*, 11 May 2020 (online).

48. Gaetano Mosca, *The Ruling Class* [ed. and intro. A. Livingston; trans. H. Kahn] (New York: McGraw-Hill, 1939 [1896]), pp. 72–73.

49. Karel van Wolferen, *The Enigma of Japanese Power: People and Politics in a Stateless Nation* (New York: Knopf, 1989).

50. Bode's seminal article is included as a chapter in the volume edited by Bob Jessop and Henk Overbeek, *Transnational Capital and Class Fractions: The Amsterdam School Perspective Reconsidered* [foreword, Gerd Junne] (London: Routledge, 2019).

51. See Nikos Poulantzas, *Pouvoir politique et classes sociales*, 2 volumes (Paris: Maspero, 1971).

52. John Maynard Keynes, *The General Theory of Employment, Interest and Money* (Basingstoke: Macmillan, 1970 [1936]), p. 376.

53. Gary Burn, *The Re-emergence of Global Finance* (Basingstoke: Palgrave Macmillan, 2006); Jacob Morris, 'The revenge of the rentier or the interest rate crisis in the United States,' *Monthly Review*, 33 (8), 1982, pp. 28–34.

54. Francis Fukuyama, 'The End of History?,' *The National Interest*, 16, 1989, pp. 3–18.

55. Cited in Leon Kreitzman, *The 24 Hour Society* (London: Profile Books, 1999), p. 19.

56. Patrick Zylberman, *Tempêtes microbiennes: Essai sur la politique de sécurité sanitaire dans le monde transatlantique* (Paris: Gallimard, 2013), p. 29.

57. Samuel P. Huntington, 'The Clash of Civilizations?,' *Foreign Affairs*, 72 (3), 1993, pp. 22–49.

58. Here one can also think of Guy Debord's prophetic *La société du spectacle* (Paris: Gallimard. 1992 [1967]).

59. Philip Zelikow, 'Thinking about Political History,' *Miller Center Report* (University of Virginia), 14 (3), 1999, pp. 5–7. On Zelikow, see my *Discipline of Western Supremacy*, Vol. III of *Modes of Foreign Relations and Political Economy* (London: Pluto Press, 2014), pp. 222, 224–26.

60. Christopher Bollyn, *The War on Terror: The Plot to Rule the Middle East* [foreword, A. Sabrosky] (n.p.: Bollyn, 2017).

61. Agamben, 'Biosecurity and Politics.'

62. William K. Carroll, *The Making of a Transnational Capitalist Class. Corporate Power in the Twenty-First Century* [with C. Carson, M. Fennema, E. Heemskerk, en J.P. Sapinski] (London: Zed Press, 2010); Kees van der Pijl and Yuliya Yurchenko, 'Neoliberal Entrenchment of North Atlantic Capital: From Corporate Self-Regulation to State Capture,' *New Political Economy*, 20 (4), 2015, pp. 495–517.

63. Phillips, *Giants: The Global Power Elites*, pp.188–92, 216–17.

64. Thierry Meyssan, 'Envers qui Emmanuel Macron est-il débiteur?,' *Voltairenet.org*, 11 December 2018 (online). Other key connections of the French president including multimillionaire Mitterrand adviser Jacques Attali. Macron's time at the Rothschild bank would seem less significant.

65. Willy Van Damme, 'Duopolie dagbladen: Belgische eenheidsworst voor Nederland,' *De Andere Krant*, 3 (6), 2020, p. 3.

66. Phillips, *Giants: The Global Power Elites*, pp. 4, 123; 'Médias français: Qui possède quoi?,' separate fact sheet with *Le Monde Diplomatique*, December 2016.

67. Media Reform Coalition, *Who Owns the UK Media?* [pdf] (London: MRC/ Goldsmiths, University of London, 2021), p. 5.

68. Barbara Boland, 'Top 50 U.S. Think Tanks Receive Over $1B from Gov, Defense Contractors,' *The American Conservative*, 14 October 2020 (online); Phillips, *Giants: The Global Power Elites*, pp. 241, 246.

69. Wolfgang Streeck, *Gekaufte Zeit: Die vertagte Krise des demokratischen Kapitalismus* (Frankfurt: Suhrkamp, 2013 [Frankfurter Adorno-Vorlesungen 2012]).

70. Peter Gowan, 'Crisis in the heartland: Consequences of the new Wall Street system,' *New Left Review* 2nd series (55), 2009, p. 9. The 'Nobel Prize for Economics' is awarded by the Swedish Central Bank with the consent of the (Norwegian) Nobel Committee and meant to present economics as a quasi-natural science.

71. Phillips, *Giants: The Global Power Elite*, p. 75.

72. Zuboff, *The Age of Surveillance Capitalism*, pp. 67–68; Christopher Rude, 'The role of financial discipline in imperial strategy,' in Leo Panitch and Martijn Konings, eds., *American Empire and the Political Economy of Global Finance* (Basingstoke: Palgrave-Macmillan, 2008), p. 211.

73. Details in Nesvetailova, *Financial Alchemy*, pp. 27–37; Nesvetailova and Ronen Palan, *Sabotage: The Business of Finance*, (n.p., Allen Lane, 2020), pp. 53–64.

74. Christos Lynteris, 'The Greek Economic Crisis as Evental Substitution,' in A. Vradis and D. Dalakoglou, eds., *Revolt and Crisis in Greece: Between a Present Yet to Pass and a Future Still to Come* (Oakland, Calif.: AK Press and London: Occupied London, 2011).

75. Rasmus, *Systemic Fragility*, pp. 221–22.

76. *FXCM*, 'Will Quantitative Easing Stop the Coronavirus Pandemic Ecnomic Fallout?,' 9 April 2020. (online).

77. Tyler Durden, '"This Could Take the Dollar Down," Alasdair Macleod Warns. "There's A Real Crisis in the Winds,"' *ZeroHedge*, 31 January 2021 (orig. Greg Hunter, *USAWatchdog*) (online).

78. Jan Fichtner and Eelke M. Heemskerk, 'The New Permanent Universal Owners: Index funds, patient capital, and the distinction between feeble and forceful stewardship,' *Economy and Society*, 49 (4), 2020, p. 494.

79. Phillips, *Giants: The Global Power Elite*, p. 71.

80. Werner Rügemer, 'Blackrock-Kapitalismus: Das neue transatlantische Finanzkartell,' *Blätter für deutsche und internationale Politik*, 81 (10), 2016, pp. 75–84.

81. Fichtner and Heemskerk, 'The New Permanent Universal Owners,' pp. 496, 510.

82. Phillips, *Giants: The Global Power Elite*, p. 51.

83. Rudolf Hilferding, *Das Finanzkapital* (Frankfurt: Europäische Verlagsanstalt, 1973 [1910]).

84. Phillips, *Giants: The Global Power Elite*, p. 155.

85. Ibid., pp. 163–82.

86. Anthony Hall, 'Was COVID-19 a Cover for an Anticipated or Planned Financial Crisis?,' *Global Research*, 20 August 2020 (online).

87. Michael Snyder, 'Why Did Hundreds Of CEOs Resign Just Before The World Started Going Absolutely Crazy?' *End of the American Dream*, 24 March 2020 (online); Radhika Desai, 'The Unexpected Reckoning. Coronavirus and Capitalism,' *Canadian Dimension*, 17 March 2020 (online).

88. Thomas Wade, 'Timeline: The Federal Resaerve Responds to the Threat of the Coronavirus,' *American Action Forum*, 29 June 2020 (online).

89. *The Conversation,* 'Paying for coronavirus will have to be like war debt—spread over generations,' 11 June 2020 (online).

90. Austin Fitts, 'The Injection Fraud—It's Not a Vaccine.'

91. *Daily Mail,* 23 May 2020, 'Lockdowns failed to alter the course of pandemic and are now destroying millions of livelihoods worldwide, JP Morgan study claims' (online).

92. Robinson, 'Global Capitalism Post-Pandemic,' p. 7.

93. Flo Osrainik, *Das Corona-Dossier,* p. 117; *Wikipedia*, 'MasterCard.'

94. Hall, 'Was COVID-19 a Cover?'

95. W. Stanley Jevons, 'Sun-Spots and Commercial Crises,' *Nature,* vol. 19, 1979, pp. 588–90.

96. Nafeez Mossadeq Ahmed, 'How Facebook Will Rule the World—Unless We Stop It,' *Insurge Intelligence*, 23 December 2017 (online); Eric Gordon and Jessica Baldwin-Philippi, 'Making a Habit Out of Engagement: How the Culture of Open Data Is Reframing Civic Life,' in Brett Goldstein (ed. with Lauren Dyson), *Beyond Transparency: Open Data and the Future of Civic Innovation* (San Francisco, Calif.: Code for America Press, 2013), p. 139.

97. *The Observer,* 'Robert Mercer: The big data billionaire waging war on mainstream media,' 26 February 2017 (online).

98. Pepe Escobar, 'Future of Western Democracy Being Played Out in Brazil,' *Consortium News*, 9 October 2018 (online); Luc Rivet, 'The Movement: What is Steve Bannon cooking up in Europa?,' *Sputnik,* 18 September 2018 (online).

99. Ahmed, 'How Facebook Will Rule the World.'

100. Ibid.; *The Observer,* 'Robert Mercer: The big data billionaire.'

101. Jörg Nowak, 'Brazil: Fascism on the Verge of Power?' *SP the Bullet*, 17 December 2018 (online); Escobar, 'Future of Western Democracy.'

102. Nowak, 'Brazil: Fascism on the Verge of Power?'

103. Christophe Guilluy, 'The *gilets jaunes* are unstoppable' (interview by Fraser Myers), *Spiked.com,* 11 January 2019 (online).

104. Zylberman, *Tempêtes microbiennes,* p. 424.

105. Thomas Ferguson, *Golden Rule: The Investment Theory of Party Competition and the Logic of Money-Driven Political Systems* (Chicago: Chicago University Press, 1995).

106. *BNN,* 'Pim Fortuyn Moordcomplot,' *YouTube* documentary, 19 May 2012 (online)

107. Stan van Houcke, 'Geert Wilders en Israël,' *Stan van Houcke blogspot,* May 2007 (online); Naomi O'Leary, 'Geert Wilders' American Connections,' *Politico,* 14 February 2017 (online); *Follow the Money,* 'Het financiële fundament onder Forum voor Democratie: Zuidas, vastgoed en orthodoxe christenen,' 14 Oktober 2020 (online).

108. Naná de Graaff and Bastiaan van Apeldoorn, 'The transnationalist U.S. foreign-policy elite in exile? A comparative network analysis of the Trump Administration,' *Global Networks,* 21 (2), 2019, pp. 229, 235–36; Bastiaan van Apeldoorn and Naná de Graaff, *American Grand Strategy and Corporate Elite Networks: The Open Door since the end of the Cold War* (London: Routledge, 2016).

109. Rachael Revesz, 'Eric Trump "said we have all the funding we need out of Russia," golf writer claims: A journalist recalled the statements from the now-President and his son during a trip to North Carolina,' *The Independent,* 7 May 2017 (online); Craig Unger, 'Trump's Russian Laundromat: How to use Trump Tower and other luxury high-rises to clean dirty money, run an international crime syndicate, and propel a failed real estate developer into the White House,' *The New Republic,* 13 July 2017 (online).

110. Wikileaks, *The Wikileaks Files: The World According to U.S. Empire* [intro. Julian Assange] (London: Verso, 2015), pp. 214–16.

111. Phillips, *Giants: The Global Power Elite,* pp. 155–58.

112. William M. Arkin, 'Exclusive: Inside the Military's Top Secret Plans if Coronavirus Cripples the Government,' *Newsweek,* 18 March 2020 (online).

113. Ibid.

114. *Cybereason Blackout,* 'Operation Blackout Summary of Events,' 5 November 2019 (online)

115. Whitney Webb, 'Engineering Contagion: UPMC, Corona-Thrax and "The Darkest Winter,"' *Unz Review,* 25 September 2020 (online).

116. Polly St. George, 'A Coup in the Making,' *Amazing Polly,* 10 September 2020 (online).

117. Ibid.

118. Rosa Brooks, 'What's the worst that could happen?' *Washington Post,* 3 September 2020 (online).

119. *AsiaTimesOnline,* 'Opinions harden against the hard men: Geopolitical chaos of their own making may yet be the unmaking of a slew of leaders, Trump included,' 10 October 2020 [separate links to different correspondents online].

120. Riccardo Corsetto, 'Conte, Leonardo SpA and the U.S. Embassy behind the Election Data Switch fraud to take out Trump: Italy and Conte played a key role in the international cyber plot to take out Trump. Leonardo Finmeccanica's

Cyber Warfare satellites are among the means used.'*Quotidiano Soveranista*, 5 January 2021 (online); partly based on testimony of former CIA station chief Bradley Johnson, 'Brad Johnson: Rome, Satellites, Servers—an Update,' 19 December 2020 (online).

121. *Times of Israel,* 'White House was reportedly not interested in the intel, but it was passed onto NATO, IDF; when it reached Israel's Health Ministry, "nothing was done,"' 16 April 2020 (online).

122. Kit Knightly, '"The Storming of the Capitol": America's Reichstag Fire? The four big lies underpinning this story show it was likely a staged event,' *OffGuardian,* 7 January 2021, and 'Prepare for the new "Domestic Terrorism Bill": The Patriot Act 2 is on the way, from the same author behind the smash-hit original,' *OffGuardian*, 8 January 2021 (online).

4.

THE VIRUS SCENARIO AS THE BASIS FOR A SEIZURE OF POWER

In this chapter we ask why the fear scenario rolled out in 2020 took the form of a virus alert and not, for example, a major terror attack such as 9/11, or a serious border incident with Russia or China.

In fact, ever since the 'end of history' at the time of the collapse of the Soviet Union, the virus route for instilling fear has been a subject of research, usually in combination with terrorism. Outbreaks of new infectious diseases have all along been regarded as a plausible reason for making people give up freedoms, not least because in many countries legislation for such an emergency has been in existence for a long time. Such outbreaks could originate by natural causes or by accidental or deliberate escapes of microbes from a laboratory (in Chapter 5, we will see that such laboratories operate in the context of U.S. military research, reaching even into China).

In hindsight one can see that a public health emergency has the best chance to bring about the consent, indeed active participation, of the population; after all, there is no formal 'opponent' facing charges, who can provide alternative information to refute them to an organized opposition. With individual governments and politics, including the media and the large Internet platforms, kept on a tight leash by the internationalized state, people who don't trust the official story are thrown back on themselves, lacking guidance, and with only social media to get their information from. In this chapter we investigate how the WHO, which had become dependent on project funding after the progressive evolution of the UN system had come under attack in the 1980s, was persuaded to change the definition of pandemic so that it could be declared earlier. The 'sleeper contracts' with the pharmaceutical industry for the supply of vaccines and medication, as laid down in the International Health Regulations in 2005, could then come into effect immediately.

After the financial collapse of 2008 the results of previous exercises and simulations were elaborated into very detailed scenarios for a possible state of emergency due to a presumed medical crisis situation. The lessons from the SARS-1 'pandemic' were essential in this regard, although mortality rates from that outbreak, as with bird flu and swine flu, remained below 1,000 and no global emergency could possibly be declared on that basis. Only in January–March 2020 did the intelligence-IT-media power bloc have sufficient grounds to unleash the information war against the population and could the state of emergency be introduced. I begin with a few comments about viruses as background.

VIRUSES AS A THREAT OR AS BUILDING BLOCKS OF LIFE?

Viruses have been around for at least 550 million years and are the most common, albeit incomplete, organisms on Earth. They are incomplete because a virus can only reproduce inside a host cell. There are an estimated 100 million different types of viruses, and they lend themselves to political fearmongering because most of us associate them primarily with disease, even with death, as in the case of AIDS. Hence, also thanks to the support of the media, the announcement of an epidemic caused by 'an unknown virus,' SARS-2, triggered existential fear and with people in a state of fear, anything is possible. We have already seen that fear had become the new political control tactic after 1989–91, now that capitalism had forfeited the ability to realize a meaningful social contract.

In reality, there are only a few hundred types of viruses that can cause disease in humans. SARS-2 has added another. As Professor Luc Montagnier, who won the Nobel Prize for Medicine for his discovery of the genome of the HIV virus, has established, this new virus was developed in a laboratory; it contains the components of HIV-1 and -2 and malaria, something not really possible in nature. As we will see in Chapter 5, several other scholars, indeed even insiders in the U.S. virology network, came to the same conclusion. However, this discovery was immediately slammed down as it was apparently judged best to keep the role of bio-warfare laboratories out of the picture.[1]

The fact that when we hear about a virus we think of disease first, is owed in large part to Louis Pasteur, who discovered the pathogenic potential of microbes (bacteria, viruses) at the end of the 19th century. His discovery was a big step forward in terms of hygiene and has saved countless lives. But when at the beginning of the twenty-first century, it came to light that the human genome too is also partly made up of viruses, this revolutionized thinking. Viruses, which may have existed before the appearance of bacteria, are now believed to have played a critical role in evolution because of their ability to mutate.[2]

The most familiar viruses are the seasonal ones that pass through the population one after the other when temperatures drop: rhino-viruses (common cold, mostly in autumn), influenza A and B (flu, winter) and corona viruses (spring). Normally, the human organism absorbs these exogenous viruses and makes them endogenous; this usually goes unnoticed—unless our immune system is weakened. In addition to a cold or illness, the assimilation of viruses also has an effect on our moods and our ability to respond to stress, among other consequences that we will leave aside here.

In Bernal's formula, life is the mode of movement of protein-nucleic acid combinations and that also applies to viruses when they 'come to life' inside the host cell of a plant, animal or human.[3] Viruses are droplets of nucleic acid packed in protein; the nucleic acid molecules form either a double helix (DNA) or a single one (RNA). According to some scholars, RNA is an even earlier life form than DNA (even before bacteria, which do not always depend on a host but live on their own). An RNA-based, so-called retrovirus (such as HIV) can be absorbed into our (DNA-based) system with the help of an enzyme, 'reverse transcriptase,' which the RNA virus brings along itself.[4] At least 43 percent of the total human genome consists of viruses and their directly derived products (proteins); about 8 percent are endogenous retroviruses (other estimates are much higher).[5]

If we understand organisms as metabolic networks overlapping with and supporting each other, we can perhaps compare viruses to the spores of mushrooms in the forest floor, which provide food for the visible plants to grow and later take it back when the plant dies. This happens not only on a chemical but also on a genetic level. In that sense, viruses supply information (genetic code), comparable to an update of a computer program. That is not just any comparison, because as we shall see, the experimental nucleic acid therapy sold to us as a 'vaccine against corona' is based explicitly on the similarities between viruses and computer updates. Such virus updates can also be 'malware' with fatal consequences, as in the HIV/AIDS epidemic, which caused more than 30 million deaths—incidentally without causing social life being shut down as in the Covid state of emergency. Even in the case of HIV, however, a symbiosis is eventually achieved between the virus and the host.[6]

The human body contains the deposit of millions of years of evolution, the microbiome, an internal ecosystem that also contains bacteria and eukaryotes (organisms with a nucleus, such as fungi and amoebas), in addition to viruses. The microbiome is part of the immune system and is kept in good condition by a healthy diet and lifestyle (and weakened by aging, disease, medication, and antibiotics). So there is a constant exchange of viruses between the microbiome and the metabolic networks of which we are a part; the outgoing ones are 'exosomes' (there is also a theory that all viruses are exosomes, inert matter being ejected and becoming reproductive again inside the cell of a new host).[7]

'Mutation and selection,' the key concepts in Darwinism, thus possibly go back to the symbiosis between a virus and a host, because when a change occurs in which a new genome is created in addition to an already existing one, natural selection comes into play at that level (the viruses that have played the largest role in evolution are remarkably similar to HIV-1, one of the components of the SARS-2 virus).[8] Because viruses constantly mutate, the cells of one organism, say, algae, can be reprogrammed by a virus into cells of another, e.g. a primitive snail. If it thereby introduces a genetic trait that the host does not already have, this can also be fatal.[9] If a population becomes exposed wholesale to totally unknown microbes (bacteria and viruses), such as happened to the indigenous peoples of Latin America after the arrival of the Spaniards, this can lead to mass extermination.[10]

Now the viruses that killed the Amerindians were transferred by humans (also via domestic animals), and their genomes were related. That also made it easy for them to be passed on between people henceforth. The HIV virus, which most likely originates in the chimpanzee and thus has origins close to humans, likewise could easily be passed on between humans.[11] If on the other hand, a virus is coming from a genetically distant organism, say, a bat, then the transmission between people is more difficult—unless it has been upgraded in a laboratory as part of so-called *gain-of-function* research as practiced under the auspices of U.S. bio-warfare institutions. I will come back to this in detail in Chapter 5.

THE JUSTIFIED SPECTER OF THE SPANISH FLU

The Spanish flu of 1918, which caused tens of millions of deaths, still haunts the public imagination as far as infectious diseases go. Tradition has it that a new flu virus hit the war-weakened population and left a deadly trail across Europe. However, the disease was neither Spanish nor a flu (since Spain as a non-belligerent had no press censorship, the first newspaper reports came from there, and 'flu' was a common term for infectious diseases). In fact, the disease started at an army base in the United States, where conscripts were called to arms to be shipped to Europe after the declaration of war in April 1917. They were obviously healthy young men. Most of the deaths among the troops, as well as dismissals due to illness, occurred in the U.S., even before their crossing.[12] Ultimately, half of all victims on both sides of the Atlantic were between the ages of 15 and 34; the over-75s, who are normally most seriously affected by influenza and again in the Covid epidemic in March-April 2020, had a lower death rate in 1918 than in the period leading up to the pandemic.[13]

Of course, a disaster causing tens of millions of victims cannot be easily traced back to a single cause, but there are many indications that the Spanish flu was the result of a vaccination experiment gone awry. Vaccination is something about which one should always be careful, given the vulnerability and individual variability of the immune system; the combination of vaccines is an especially risky business. Virtually all vaccines are grown on animal tissue, e.g. mouse brains, monkey kidneys, etc. A vaccine can therefore reactivate endogenous viruses in humans and render them pathogenic again by introducing new, exogenous ones. For Covid-19, a new, revolutionary injectable serum has been developed which is aimed at making our cells produce the virus protein against which our immune system will then mobilize. The consequences of this type of injection, which falsely uses the legal possibilities created for vaccines, may be unforeseen. However, the interests at stake for the pharmaceutical industry and the institutional context in which it operates, such as the Gates Foundation, the WHO and other parts of the biopolitical complex, are enormous. Hence a definite nervousness concerning the true nature of the Spanish flu itself has arisen again. Negative publicity about vaccination is the last thing the interested parties are waiting for, and the claim that vaccination may have been the cause of the Spanish flu pandemic of 100 years ago, has to be refuted once again today. However, the counterclaims are anything but convincing. Thus, a recent Reuters fact-check provides as counterarguments: 1) the vaccination explanation is wrong; 2) it wasn't the first time that U.S. soldiers were compulsorily vaccinated, and 3) more soldiers died of flu and pneumonia in that war than under enemy fire.[14] Since nobody denies 2) and 3), and 1) can be left aside, the question remains, what really happened?

The story begins with the Rockefeller Foundation, established in 1913. By its establishment the founder of the Standard Oil empire converted his wealth into social and political power beyond the reach of the taxman. Medicine was Rockefeller's primary interest and he early on added his weight to the promotion of allopathy, the approach to disease based on the notion it must be fought from the outside. This method comprises targeted prevention via vaccination and synthetic medication that tackles the specific condition. The alternative approach is the holistic one, which includes natural remedies aimed at the whole human being and also homeopathy, which assumes that the body itself has to overcome a disturbance. Coal tar, a by-product of the petroleum industry, became the basis for synthetic medicines and the more influence of the wealthy Rockefeller institutions on medical education and treatment methods, the more allopathy.[15]

Since drugs have become the third leading cause of death in the West after cardiovascular disease and cancer,[16] this approach should raise questions. But allopathic medical science is not only medical. The Rockefeller Foundation interpreted the improvement of education, hygiene and sanitation, as well as promoting the control of human behavior, as steps towards making societies not only happier, but also closer to the American model. This in turn increased receptivity to the political influence and capital of the United States. 'Directors of the Rockefeller Foundation wanted to both improve the plight of local populations and make the world a more hospitable place for North American elites,' Ilana Löwy and Patrick Zylberman write.[17] But their influence on the practice of Western medicine has resulted in alternative remedies, from Chinese acupuncture to homeopathy, being labeled quackery and dismissed even today, rather than ranking them according to proven effectiveness. Yet these latter remedies have an unforeseen characteristic in common with modern quantum physics, namely, they start from the whole, while the older, Newtonian mechanics is reductionist; matter can be broken down into elements, but waves (the unit of quantum physics) cannot.[18]

After the declaration of war in April 1917 the Rockefeller Institute for Medical Research in New York (today, Rockefeller University) was authorized to begin vaccinating soldiers about to be shipped to Europe with an experimental bacterial serum against meningitis, grown on horses. Whether a Rockefeller mission that secretly traveled ahead to France in 1917 had any connection with this project, is not known.[19] The vaccinations took place at Fort Riley Army Base, Kansas, between January and June, 1918. From there the journey went by train to the Atlantic coast and onwards by ship, under less-than-ideal sanitary conditions. In many cases this led to bacterial pneumonia. Once in Europe, according to official figures, for every 1,000 men, 977 were hospitalized; then the deadly infection also spread among the civilian population, and as mentioned, it affected not the typical risk groups for influenza, i.e. the elderly and people with underlying diseases, but children and young people. Compulsory face masks made the disaster complete. In the meantime, the Rockefeller Institute got so many orders for the serum from allied countries that an extra horse stable was put into use to be able to produce enough of it. When the military were slated to return to the United States after the unexpected cessation of hostilities, residual supplies of various vaccines were injected into them and, once back, they infected millions of others.[20]

In 2005 the genetic sequence of the virus said to have caused the Spanish flu was published (in fact, it was a laboratory reconstruction). In addition to the bacterial infection, the pandemic involved a series of childhood diseases such as measles and diphtheria, which under normal circumstances would never have affected healthy young men. In 2008, a paper from the National Institute of Health (one of the authors was Dr Anthony Fauci, director of the NIH and of the National Institute of Allergies and Infectious

Diseases, NIAID) found that in 92.7 percent of the 9,000 autopsies that had been investigated anew, the cause of death was bacterial pneumonia.[21]

BIOTERROR SCENARIOS AND THE SARS OUTBREAK

Among the medical worst-case scenarios governments began to study in the 1990s with an eye to upholding their authority despite the moral disintegration of capitalist society, the Spanish flu held pride of place. By the time of its centennial anniversary in 2018, the world had simulated and/or exercised the arrival of new epidemics for decades. For after the biopolitical complex initially declared the era of infectious diseases closed (in the late 1960s), alarming reports of their return began coming in during the 1990s. The collapse of the Soviet bloc and of many other countries that owed their sovereignty to its existence, as well as the wars that erupted as a result, led to sharp declines in their living standards and new emigration flows. The migrants brought the diseases with them from the disaster areas. The WHO might have closed the tuberculosis division in 1989, but four years later, reports came in that the disease was back; laboratories were now instructed to resume their search for bacteria and viruses.[22]

In hindsight, infectious diseases and bioterrorism received exceptional attention among the fictitious threats for which scenarios have been developed since the promulgation of the 'end of history.' Bioterrorism is the ideal threat, because it only takes one person to cause a disaster. Under these circumstances, health was increasingly associated with national security. In May 1989, a conference entitled 'Emerging viruses' was held in Washington with the support of Fauci's NIAID and Rockefeller University. Also in 1989, the Johns Hopkins Center for Civilian Defense Studies enacted a scenario where Washington DC was hit by a smallpox virus attack. According to Zylberman, politics as a result descended into the realm of fiction. This was quite literally the case when President Bill Clinton read a science-fiction novel about a virus attack on New York, *The Cobra Event*, and asked a Pentagon deputy secretary of defense if this was really possible. Since the official in question had also advised the author of *The Cobra Event*, the answer was 'yes.'[23] The various medical institutes of Johns Hopkins, which, as we will see, are the key academic backers of the Covid state of emergency, have a long history of outside funding, starting from the Rockefeller Foundation in 1916 to the record subsidies (one of $1.8 billion) by Michael Bloomberg, for the Bloomberg School of Public Health, among others [24] While there had certainly been terror incidents—the 1993 attack on the World Trade Center, a poison gas attack in the Tokyo underground in March 1995, and the Oklahoma City bombing a month later—bioterrorism was the one thing that had never happened prior to Clinton's $3 billion budget approved by Congress to fight it, passed in January 1998. Zylberman's conclusion is that epidemics are being politicized and terrorism is becoming medicalized, all on the basis of a low probability, but potentially enormous consequences of their occurrence. In the words of Catherine Austin Fitts, epidemics have become the medical variant of 'false flag' operations.[25]

Looking back today it is striking how scenarios for bioterrorism all had as their premise that in the panic following a biological attack it would be possible to make people give up their freedoms for the sake of the fight against the epidemic and agree that dissenting voices should be suppressed. On that basis a convergence came about between epidemiologists and those looking for ways to reverse civil liberties and roll back democracy. The institutions that still play a major role in the Covid crisis today were involved right from the start. A state of emergency will always involve thousands of specialized personnel who are kept in readiness for it and who will in good faith execute their tasks accordingly. As the example of 9/11 illustrates, their number can even be invoked to contradict the idea of a contrived crisis or false flag incident. The Johns Hopkins Center for Civilian Biodefense Studies was also established in 1998 and symposiums on bioterrorism were held in the following year and in 2000, with politics, science and media becoming increasingly intertwined. Matters previously left to experts were now made part of the broader political process, amplified by alarming press reports. The first exercise against bioterror was also held that year (*TopOff2000*), and several others would follow. The Johns Hopkins Center produced two scenarios for a bioterror attack, Dark Winter (practiced in June 2001) and Atlantic Storm (January 2005).[26]

Dark Winter was held at Andrews Air Force Base in Camp Springs, Maryland. The name Dark Winter came from Robert Kadlec, an intelligence lobbyist who had already worked for the Bush Sr. administration. In 2019 he would lead the Crimson Contagion exercise and in 2020 he was tasked with the Covid response in the United States. The scenario was based on the story line that Saddam Hussein had obtained the smallpox virus from specialists in the former Soviet Union (Dark Winter was one of the sources of the claim that Iraq possessed weapons of mass destruction). Upon its introduction into the U.S., an outbreak occurred in Oklahoma and still according to the script, panic broke out. Several members of Congress, a former CIA director (James Woolsey), as well as government and trusted media representatives (including Judith Miller of *The New York Times*) participated, playing the role of the National Security Council. The script was written by the Johns Hopkins Center and the Institute for Homeland Security ANSER, and performed by several more institutions, including the Center for Strategic and International Studies, (CSIS) of Georgetown University in Washington.

Prominent Neocons (Richard Perle, Donald Kagan) meanwhile predicted with great certainty that a real bio-terror attack was imminent; Vice-President Cheney took anthrax medication after being briefed by Tara O'Toole and other members of the team that set up Dark Winter. How the participants in Dark Winter could uncannily anticipate an anthrax attack, we don't know. Yet very soon thereafter, an actual anthrax attack took place on U.S. soil shortly after 9/11, when letters containing anthrax were sent to Senators Tom Daschle and Patrick Leahy, among others, leading to the temporary shutdown of Congress—attacks which were then attributed (by Woolsey and Israeli sources) to Saddam Hussein but were actually homegrown, with the anthrax found to have originated in the U.S.'s own biowarfare lab at Fort. Detrick. The exercise made it clear that politicians, if sufficiently scared, would view the world through the lenses provided to them by experts.[27] The response to the actual attack reinforced that message. It included vaccinating U.S. military personnel against anthrax before their being sent to Afghanistan in Iraq in the 'War on Terror' and in spite of the resultant deaths and serious side-effects, that program was resumed in 2006 after having been interrupted on a judge's order.[28]

Dark Winter already warned against disinformation about medications and other matters endangering public safety. For the suspension of civil liberties, it had recommended the proclamation of the state of emergency and that the president should also use his prerogatives under the Insurrection Act. That Act, as we saw, allows military rule to be installed, and entails suspension of the freedom of assembly, prohibition of travel, imposition of quarantine, and even overrides *habeas corpus* (physical inviolability of the person). Military courts should take over in case the civil justice system can no longer handle the situation. Both the TopOff exercise and Dark Winter took into account the possibility that the public and doctors would oppose the state of emergency and that coercion would be necessary to force them into submission. For Vice President Cheney the exercises were sufficient ground to argue for mandatory universal vaccination—a chilling prediction of what has become familiar to us all in the Covid crisis, twenty years later.[29]

Although it was recognized that overly draconian measures might drive public opposition underground, thus exacerbating a real pandemic, a law was passed in the U.S. in 2002 (the Public Health Security and Bioterrorism Preparedness and Response Act) which increased the power of the government and gave the police the authority to put people without symptoms into quarantine. In France, too, the possibility of detaining asymptomatic persons was established (in a decree of 2003).[30] In late 2002, the

White House ordered, on the pretext of an imminent bio-terrorist attack, and against medical scientific advice, that half a million Americans should be vaccinated against smallpox. A Citizen Corps of vigilantes was formed to make America a safer place; in 2006, 2,000 such vigilante teams were already active.[31]

Meanwhile in November 2002, the original SARS, Severe Acute Respiratory Syndrome, had broken out in Guangdong, China. The virus was said to have originated in a market where live animals were sold for slaughter. The Dutch celebrity Professor, Ab Osterhaus of the Erasmus Medical Center in Rotterdam, confirmed at the request of the European Commission that the source of the outbreak was a new coronavirus from bats. Another virologist whose name would come back later, Dr Peter Daszak, came to the same conclusion.[32] However, the French professor of pharmacy, Jean-Bernard Fourtillan, who, as we saw, was forcibly detained in a mental institution in December 2020, released documents showing that the Institut Pasteur had patented the SARS-1 coronavirus in 2003; it contained components of malaria, which cannot be present naturally, but is the result is gain-of-function research in a laboratory, an activity undertaken with the stated intention of developing a vaccine against it—or for some other reason. Patents, moreover, cannot be granted for a natural substance or process. In 2011, the Institut Pasteur would also patent a SARS-2 because the SARS-1 patent would expire in 2023.[33] This explains, incidentally, why on social media documents have been circulating which show that long before the outbreak of SARS-2, there was a (patented) medication against it. We will come back to this later.

The transmission of a virus from animals to humans, *zoönosis,* was supposedly at the origin of both SARS outbreaks. This has been attributed to humans pressing forward into the habitat of wild animals; enabling viruses to then pass through contact, food or air (sometimes with domestic animals as a stopover). The bird flu virus, originating in aquatic birds, upon mutation into H5N1 jumped from chickens to humans and proved fatal for them. Ultimately, however, transmission between people turned out to be very limited. 'Habitat compression' (in addition to pollution, rising temperatures and other forms of change of biotope) has also become the explanation for the Covid crisis of the 'responsible left,' which since 9/11 has joined the blanket rejection of 'conspiracy theories.' For example, according to Sonia Shah in both *The Nation* and *Le Monde Diplomatique*, HIV, Ebola in West Africa, and Zika in the Western Hemisphere are examples of zoönosis due to habitat compression; the connection with recent clearing of jungles and the like is demonstrable.[34]

However, as noted, the transmission of such a virus between humans is limited by the difference in genome; the further removed the animal is from humans, the less chance of human–to–human infection. According to Dr Judy Mikovits, there is certainly transmission from animals to humans, but then it was transmitted via vaccination, as in the cases of HIV and Ebola. HIV is said to have arisen when an oral polio vaccine was administered in the Belgian Congo in the 1950s, which was grown on kidney tissue from chimpanzees and bonobos. After millions of people in Congo and Rwanda-Burundi had received the vaccine, the fatal growth of HIV began. HIV has subsequently killed 30 million by AIDS and by 2010 had infected an estimated 34 million people.[35]

The exercises and simulations of bioterrorist attacks continued in the years after 9/11, becoming more extensive and expensive. The 2005 Atlantic Storm scenario projected a shocking level of preparation among the NATO allies, but then, a bioterror attack had never happened. In Atlantic Storm, former Secretary of State Madeline Albright played the president of the U.S., while Bernard Kouchner, the founder of Médecins Sans Frontières and future minister of foreign affairs, played the president of France; Gro Harlem Brundtland played her own former self as head of the WHO. The TopOff3 scenario also took place in 2005 and was headed by Michael Chertoff, Secretary of Homeland Security. TopOff4 went a step further in 2007 in terms of participants and organizations involved. The costs had also risen, from $3 million in 2000 to $25 million in 2007. The exercises were internally

internally criticized for fueling a climate of fear even though the exercises were secret, not intended for public consumption.[36]

By now the concerns surrounding implementing a biopolitical state of emergency had been worked out in detail.

THE WHO AND BIOPOLITICAL SOVEREIGNTY

The Chinese government withheld information about the SARS virus for months in 2002–03 and did not notify the WHO. This allowed the infection to spread. Modern forms of travel such as flying have greatly increased the likelihood of an epidemic, so withholding information was a serious matter. That is why the WHO took on a larger, political role in response to the SARS crisis. But this had been preceded by a serious crisis at both the WHO and the UN system as a whole.

In the 1960s and 1970s, the United Nations had begun to reflect the concerns over their economic sovereignty of a large number of new states that had acquired formal independence by decolonization and joined the UN as sovereign nations. Many of the new states had discovered, however, that political independence as such did not necessarily imply material sovereignty. If a country's resources are still being exploited by foreign corporations, neither flag nor national anthem are sufficient to ensure the population's well-being and ultimately, its loyalty. That is why in the course of the 1960s, a movement for a New International Economic Order (NIEO) that should make the decolonized countries economically sovereign too, was gaining momentum. Elsewhere I have described how such a UN-coordinated world economy restricted the freedom of movement of capital and hence, was completely unacceptable to the West. Neoliberalism in many ways was a reaction to the NIEO, just as it had developed in response to other progressive trends of the period.[37]

The campaign against the UN was spearheaded by the Heritage Foundation in the United States and by the lesser known, but highly influential International Chamber of Commerce (ICC) in Paris. Under heavy pressure from these quarters, the Egyptian UN Secretary-General since 1992, Boutros Boutros-Ghali, was first compelled to shut down their much-hated UN Center for Transnational Corporations (CTC) and several other UN programs. The CTC report on pollution by multinationals, meant for the UNCED environmental conference at Rio de Janeiro, never arrived there. When in early 1996 Boutros-Ghali proposed to introduce a global tax to overcome the UN's pressing financial crisis, Washington gave short shrift to his ambition for a second term in office. Kofi Annan, a Ghanaian alumnus of MIT's business school, succeeded him. Shortly after taking office, Annan signed an agreement with the WEF in Davos linking the UN Secretariat to an electronic communications system, WELCOM, that put him and his staff in direct contact with a number of government leaders and with the directors of the World Bank and other international financial institutions. Progressive NGOs, which had played an important role in the NIEO orientation of the UN system in the preceding period, were put at a distance. Less than six months later, a summit meeting on the role of the business world in the political development of the UN, followed. In addition to Annan, it was attended by B. Stigson of the Business Council for Sustainable Development (BCSD), Maria Livanos Cattaui (Secretary General of the ICC) and some twenty CEOs and government representatives including Larry Summers, U.S. Deputy Secretary of the Treasury.[38]

The UN system was now largely subordinated to the Western bloc and transnational corporations. In 1999 Annan presented a plan for sustainable cooperation with the business world, in particular the ICC: the Global Compact. This Compact aimed to protect the rapidly advancing internationalization of the economy from friction by incorporating certain minimum standards in the areas of industrial relations and environmental protection. As we will see, the World Economic Forum's Great Reset is inspired by the same approach, not to carry out a diabolical 'green' global project, but to ensure that the continuing internationalization of capital will not be derailed by a workers' or environmental movement militating against the excessive exploitation of society and nature. Even so, on the same day that the Global Compact was announced, Livanos Cattaui wrote an Op-Ed piece in a U.S. newspaper opining that all this should not be read as meaning that companies would henceforth be judged by outsiders. Annan therefore limited himself to publishing the names of companies that excel in the domains identified, thus establishing the standard of *best practice*.[39]

The WHO had also been targeted in the campaign against the UN, notably by Nestlé of Switzerland, a company prominently represented in the Atlantic networks of joint directorates and private planning groups. Nestlé in particular fulminated against the WHO because the health organization wanted to reduce the use of synthetic baby milk in the Third World; the WHO also encouraged poor countries to produce their own medicines, which provoked the big pharmaceutical interests.[40] The major American tobacco producers, too, had the WHO in their sights on account of its anti-smoking campaign. With sponsored publications by 'experts' from academia and the media, the tobacco interests were able to discredit the WHO as a 'socialist' bulwark.[41]

As a result of these attacks, the WHO became more and more dependent on non-budgetary funding, i.e., money from governments and companies attaching specific conditions to their donations. The WHO's total budget in 1996 was less than the loans for health purposes provided by the World Bank. The Bill & Melinda Gates Foundation then stepped in, offering a helping hand from 2000 onwards. As of that year, the foundation donated a total of $2.5 billion to WHO (about 5 percent of its $52 billion foundation assets; figures up to 2017). In 2004, public-private partnerships were added, giving the pharmaceutical industry direct access to WHO decision-making.[42]

In the meantime, in 1998, in order to restore the WHO's effective role in the changed relationships, the aforementioned former Norwegian prime minister, Gro Harlem Brundtland had been made Director General. Under her leadership, the health organization began to make use of legally binding (treaty) instructions. Paradoxically, the first success was a Framework Convention for tobacco policy. Such conventions provide a basis on which negotiations on more detailed regulations can then be held.[43] This restored WHO prestige and the Global Outbreak Alert and Response Network, intended to enable rapid response to epidemics, could for example be brought back in its remit (although there was still debate whether that task would be better entrusted to the UN, the EU or even to NATO).[44] Yet, the sovereignty that the WHO thus acquired remained a *derived* sovereignty. Because of its new financial dependence, going forward it came, as did the new neoliberal world order as a whole, under the dominion of capital, implemented by the internationalized state.[45]

After a period in which infectious diseases and in particular, influenza, had received less attention within the WHO, the general change of mood following 9/11 and the anthrax letters to the two senators and some media, and then the 2002–03 SARS-1 epidemic, worked to propel the health organization step by step to the center ground again. Towards early 2002 it recommended that governments begin monitoring their inventories of available smallpox vaccine, and prepare for an emergency.[46] During SARS, the health organization recommended that all non-essential travel to Hong Kong, Toronto, Beijing, and the Chinese provinces of Guangdong and Shanxi be suspended, which in May 2003 was broadened to cover all of China as well as the Taiwanese capital, Taipei. But although signs of spread to the Canadian province of Ontario had been reported in June (as we will see, the experience of the provincial capital, Toronto, would provide the template for future lockdowns), in July it was made official: the spread had stopped.[47] Even so, Margaret Chan, the Director-General of WHO, originally from Hong Kong, was criticized for her passive attitude and under pressure from U.S. institutions (CDC, USAID) and international bodies in which the U.S. plays the leading role (World Bank, G7), the WHO embarked on a more proactive stance on infectious diseases, now stipulating that this could take the form of quarantines as well as vaccination, while also tackling HIV/

AIDS more vigorously. The concept on which this rested was the notion of human security, specifically biosecurity.[48]

The WHO recommendation to governments to stock up on influenza drugs such as Tamiflu can be traced back to the 'flu interests' which during the SARS outbreak were trying to create panic by evoking the specter of the Spanish flu. The European Working Group on Influenza (ESWI), which is funded by ten major pharmaceutical companies, including GlaxoSmithKline and Novartis, wanted obligations to be imposed on countries to maintain substantial stocks of medication and vaccines (which they themselves happen to produce). Their ally in this endeavor was aforementioned Ab Osterhaus, the chairman of ESWI, who during the SARS panic was constantly featured in the Dutch media and succeeded by his fear-mongering interventions to make the Netherlands the entry point for European epidemic policy henceforth.[49]

THE INTERNATIONAL HEALTH REGULATIONS

As we saw, worst case scenarios had increasingly become a guideline for policy in the period from 1989–91. When in the aftermath of SARS-1, a new variant of bird flu, H5N1, was discovered, which especially affected vulnerable poultry and which by 2005 proved to be able to jump to humans, this prompted the WHO to take steps towards a transfer of national sovereignty to the WHO in the name of biosecurity.[50]

This important development (which, as argued above, actually amounted to a transfer of sovereignty to an internationalized global state dominated by international capital) was laid down in the third edition of the International Health Regulations of 2005. This document originally dated from 1969, but now became binding. In these regulations, the WHO included a list of measures that states are obliged to take in the event of an outbreak. On the eve of the Covid crisis, in July 2019, the WHO, together with the World Bank, explicitly reminded member states of this obligation. What

was new was that the measures would no longer be limited to specific diseases or the mode of infection but would cover *any* disease
or medical condition that could cause significant harm to humans,
regardless of origin.[51] This extension was said to ensure that medical emergencies could not get out of hand due to slow reporting,
as in the case of China with SARS. New diseases would also come
into the picture sooner as a result of these mandatory provisions,
if necessary through unofficial channels if governments failed to
report them; states could then be asked for confirmation.

Among the measures that states would be obliged to take
going forward, we see all the elements that are also in force today:
issuing certificates for international travel and transportation, as
well as facilities for transshipment and traffic through international ports, airports and border crossings among them. In addition,
states committed themselves to impose quarantines, prescribe
mandatory vaccinations or other forms of prevention, along with
contact registration and related measures.[52]

By 2007, 178 countries had reached the point where they had
drawn up national plans to prepare for a flu pandemic. In practice,
there was little prospect that there would be sufficient vaccine
supplies in any country in an emergency, but consultations had
been initiated with the pharmaceutical industry to provide for
this through pre-contracts (so-called 'sleeper contracts') to come
into effect once the WHO declared a pandemic.[53] That the WHO
had now become a more important political body than heretofore
(albeit on the leash of other interests) is evident from the fact that
at the time of the bird flu panic in 2006 (panic was now the normal reaction as soon as something happens), the French minister
of health proposed to give the WHO powers like those of the
International Atomic Energy Agency—unlimited access to and in-
spections of places where it is suspected that something is amiss.
Thus, Zylberman comments, the minister expressed the concern
of the rich countries towards states that failed in this regard. With
regard to SARS, that was China.[54]

However, the WHO's new power was derivative—not only
through non-budgetary, conditional (project) funding, but also

because certain NGOs such as Médecins Sans Frontières (whose founder as we saw participated in Atlantic Storm and who later became François Hollande's minister of foreign affairs) took over certain tasks from the WHO.[55] In other words, they were now ready for the crises to come. Until then, predictions about new pandemics had not really come true: SARS-1, 2002–3, 774 deaths; bird flu (H5N1, Influenza A), as of 2004, despite predictions of millions of deaths, 282.[56]

THE MEXICAN OR SWINE FLU SCARE

Now that the WHO had gained in authority and its International Health Regulations had imposed binding obligations on the member states that would come into effect as soon as the storm flag was hoisted, a major lowering of the threshold for WHO intervention was implemented. On 4 May 2009, a few weeks before the outbreak of the Mexican or swine flu (H1N1, Influenza A), the health organization changed the definition of a pandemic. To be classified as a pandemic, an epidemic no longer had to meet a series of conditions such as rapid spread in many countries, no or insufficient immunity and extreme numbers of deaths or serious illness. All it took from now on was the spread of a new virus and more than normal numbers of sick people—and of course a WHO decision to declare the pandemic.

It would later turn out that the change of definition had in fact sparked heated debates in the WHO Scientific Advisory Council (SAGE). When after the promulgation of the swine flu pandemic, many millions' worth of vaccines and medicines were purchased in accordance with the provisions of 2005, but ultimately few deaths were to be mourned, Dr. Wolfgang Wodarg, a German lung consultant, in his capacity as chairman of the parliamentary assembly of the Council of Europe, called for an investigation of possible conflicts of interest in the response to the swine flu. Given that both the pharmaceutical industry advocacy group (IFPMA) and the vaccine manufacturers' organization for the developing world (DCVMN) had co-authored the plan on which the change of definition had been based, this was a most appropriate call. In an interview with *Der Spiegel*, the epidemiologist, Tom Jefferson, confirmed that the change of definition had created a huge market for H1N1 drugs.[57]

The swine flu episode also marked the first attempts to implement political-economic restructuring. In the wake of the 2008 financial collapse, the UN and the World Economic Forum had begun planning a route to recovery, the first version of what has now become the *Great Reset*.[58] This plan, the *Global Redesign Initiative*, which has since been removed from the public eye, had all the elements of the oligarchy's current seizure of power. But the biopolitical complex at that point was apparently not yet sufficiently integrated with the WEF plans, although UN Secretary-General Ban Ki-moon was present when a delegation of thirty pharmaceutical companies visited WHO headquarters in Geneva on 19 May to consult with Ms. Chan. Subsequently, on 11 June 2009, the WHO announced a 'Phase 6 Pandemic Emergency,' thereby activating the sleeper contracts (which include the provision that vaccine manufacturers are granted immunity from liability claims in case of side-effects). In addition, the effect of declaring a pandemic is to relegate vaccine safety to second place; the same would happen in the Covid crisis in 2020–21 because of the alleged emergency situation.[59]

The fear campaign in the media was another similarity between the swine flu episode and the Covid crisis eleven years later. Indeed, the generation of fear is in the media interest because frightened people 'watch more often and for longer.' In addition, even then, there were vaccines produced in haste. An increased risk in 2009 was the addition of *adjuvantia*, additives to reduce the amount of actual vaccine, to be used under a new patent. One of them is squalene, oil from the liver of sharks; so the fact that these fish are being slaughtered is not, or not only, due to the demand for shark fin soup in Hong Kong. GlaxoSmithKline, which would sell 4.4 billion euros of swine flu vaccines alone, in fact did not have a vaccine ready when the pandemic was declared, let alone

one that could be subjected to the minimum test period established by the European Medicines Agency (EMA), normally 18 to 24 months. That is why a mock-up of an existing vaccine was made. This was then tested, in this case by Osterhaus, who announced the positive result at a press conference of ESWI, which he chairs. Subsequently, a provisional approval was granted ('prequalifica-tion'), allowing GSK and Novartis to start supplying vaccines for the inoculations.

In the end, states purchased millions of vaccines but due to the limited impact of the Mexican flu, many had to be dumped. In the Netherlands 20 million doses were thrown away; in Germany, which had purchased 34 million vaccines under a secret contract with GlaxoSmithKline, 86 percent went into the garbage can.[60] And so on, in other countries. Not only that, the shots that had been given caused serious side-effects. The manufacturers were not held accountable, and neither were governments, because public health legislation in many countries where vaccination is voluntary, holds that citizens are therefore responsible them-selves. The side-effects of Mexican/swine flu vaccines included the muscle disease, Guillain-Barré, but also stillborn children. Narcolepsy, too, was frequent, which for Finland for instance was the reason it stopped the vaccination campaign altogether in August 2010. Another important similarity between swine flu and SARS-2/Covid is that in retrospect, the deaths of many patients were wrongly attributed to the disease. In reality the deceased in many cases suffered from an underlying ailment and died with, not of, the flu. The theory of a 'second wave' of swine flu also circulated in 2009–10, but at the time experts took it with a grain of salt.[61]

In the end, swine flu was not 'controlled' or 'defeated'; it *passed*, like all other viral infections, but the proclamation of the 'pandemic' effectively served as a dress rehearsal for the Covid crisis. However, much of what is now put in practice to keep the world's population under the spell of SARS-2 still needed further research before an effective scenario could be drawn up. The first and in many respects decisive research was that of the leading French sociologist of public health, Professor Patrick Zylberman.

ZYLBERMAN: MOST CITIZENS DEMAND A LOCKDOWN

Based on the experiences with the SARS-1 epidemic in China and Canada and in particular on the lockdowns that had been instituted there on the advice of the WHO (but were not yet binding), Patrick Zylberman made an extensive study of their social-psychological consequences. In doing so, he also provided the research enabling the European and other governments to weigh the possibilities of using a medical emergency to restore discipline over the population. In many ways Zylberman's work is reminiscent of, for example, Ulrich Beck's theory of risk society; it has an intelligent and critical tone, but in the end it reads as a justification of the 'new normal' in Zylberman's case, of 'the measures,' that is, the lockdowns.[62]

Zylberman spent six months as a visiting researcher at the Munk Center for International Studies and the Munk School of Global Affairs, both located at the University of Toronto, in the city where SARS had the greatest impact outside China. Hungarian-born businessman Peter Munk, the owner of Barrick Gold (the world's largest gold mining company) has funded these centers. How Barrick is embedded in the transnational ruling class is illustrated, for example, by director Dambisa Moyo from mineral-rich Zambia, who also sits on the boards of Chevron and Barclays, is a member of the Bilderberg group, the WEF, etc.[63] These are the channels, in this case reaching deep into black Africa, through which the consensus that lockdown can be an effective response to global unrest comes about, a consensus that becomes global in turn.

From 2007–09, before it came out as a book in 2013, Zylberman's work, *Tempêtes microbiennes* (Microbial storms), circulated in the form of papers for the Toronto-based Lupina Foundation (which supports health research). Janice Stein was then director of both the Munk School (until 2014) and (still today) the Lupina Foundation; she is an influential figure in North American academic circles. In 2011 and '12 the present author was able to ascertain how these circuits work in meetings at the universities of Toronto and Princeton, respectively, with Professor Stein present in both cases. At Princeton, representatives from the Council on Foreign Relations and the U.S. State Department were also there, while the Ford Foundation, the Carnegie Endowment, the German Marshall Fund, and other agencies were sponsoring that particular event.[64] Thus the insights produced in such meetings become part of the knowledge base of these and other institutions, and the same is true of Zylberman's work. That 'health' in the new millennium has been integrated into the international strategy of the ruling class, also transpires from the fact that in 2004 the first chair in health and globalization was created at the Council on Foreign Relations, funded by the Bill & Melinda Gates Foundation.

Zylberman's investigations of the lockdowns in the 2003 SARS crisis are little known to the public but have proved of great importance to the handling of the Covid crisis. The measures taken in China and Toronto, such as quarantine for people without symptoms and the tracing of contact histories, according to one WHO expert, were 'medieval'; but as it turned out, the large majority accepted them all. SARS-1 made it clear that a population can be well controlled in the event of an epidemic. The model for such a drastic regime was the Spanish flu of 1918, not the flu epidemics of 1958 and 1968. For despite killing one and a half million victims each in the latter two cases, these influenza episodes left the lives of the rest of the population untouched. To 'work' as the justification for a state of emergency, an epidemic must completely disrupt social life, bringing the state to the brink of collapse, and must shatter the morale of the population; in short, there has to be an infection that destroys everything, the social structure, institutions and customs.[65] We are in the midst of such a crisis today, officially in the name of 'the virus,' but in fact as a result of the lockdowns impacting social and economic life, still in place, and argued as likely to remain necessary or to be reimposed with the advent of the variants, even though to date the actual virus has turned out to be not much worse than a serious influenza epidemic.

The previously mentioned exercises and simulations based on scenarios looked at behavioral responses that cannot be calculated in advance with the help of models. Zylberman's studies instead aimed at finding out how collective psychology and the dynamics of social relations work in a serious health crisis and this required a different approach. The complexity of human responses and their spontaneous reflexes in a population subject to severe restrictions, must be looked at in the process itself, or in comparable situations. Even the 2005 video game, World of Warcraft, about an epidemic that affected 4 million people, afforded insights into the players' emotional responses.

Zylberman's wide-ranging investigations led him to conclude that the state is able to exercise much greater power than had been assumed in the era of privatization and liberalization. In the SARS crisis, a successful disciplining of the population proved possible, and comparatively easy at that. For whilst quarantine and school closures were not necessary at all (people without symptoms could not transmit the virus, and children could not either), voluntary confinement was nevertheless accepted. There was a new force at work, according to Zylberman, a new fiction really, and that was extreme civic spirit, patriotism even. Certainly, people in quarantine suffer from loneliness, from being cut off from the world, from boredom and fear. The longer the lockdown lasted, the more depression and stress, plus the overriding fear of losing work and income. And yet despite all that, it was accepted that face masks had to be worn, even at home. The often-contradictory instructions by the authorities aggravated frustration and unrest, and yet people followed them widely.[66] Could this have escaped the attention of those contemplating how to address global popular unrest and/or in charge of the current state of emergency?

In the SARS epidemic, health workers were much admired, just as there was great admiration for police and firefighters on 9/11. It was the trust placed in these categories of frontline workers,

much more than trust in governments that were constantly contradicting themselves, that made people willing to obey, again out
of a pervasive 'civic duty.' Zylberman calculates that 50 to 68
percent obeyed; only around 15 percent were dissidents. These
percentages applied to both sides of the Atlantic in the situations
that can be compared e.g. in France in the case of bird flu, and
probably also in the current Covid state of emergency.

Measures to deal with epidemics had been in place long
before. British legislation had already provided for hospital detention at the end of the 19th century and in 1936 this did not even
require a court decision anymore. From 1984 it was no longer
necessary to prove that someone was infected! Of course, this
was in violation of the European Convention on Human Rights.
After 9/11, new provisions were also introduced in the U.S. A
state governor could henceforth declare a state of emergency on
health grounds and the health authorities were given far greater
powers. Opposition was a crime from then on. The idea is that
the protection of public health abolishes all rights of privacy and
civil rights. Again and again the specter of a bioterrorist attack
was invoked and the actions the authorities could resort to were
not limited to quarantine, compulsory treatment and isolation, but
also intervention of the army and media censorship. All of these
measures were enshrined in the 2001 National Emergencies Act
in the United States and corresponding powers of the police and
National Guard were enacted too. While Zylberman argues that
this enables protection of public health to become pure tyranny,[67]
he has also demonstrated that if the aim is to impose tyranny,
public health has turned out to be a most effective justification to
achieve that without having it be called tyranny.

Yet there is always a risk of riots. For civic duty is a dou-
ble-edged sword: citizens obey because they are advised it is
in their own interest, but in fact they don't want to. Therefore,
restrictions must be promoted effectively. Dissenting views, indeed even rumors that might undermine confidence in the health
authorities, must be actively combated. There can also arise a dangerous disproportion between unbending governmental measures

and the more nuanced views of experts, which are based on actual facts and numbers and are not necessarily aimed at perpetuating the panic. The longer it takes, the more the nuanced experts will be listened to, is Zylberman's conclusion.[68]

The similarities with the current Covid crisis are striking. The SARS-1 epidemic, the key event for Zylberman, showed what is possible if a national state of emergency is to be maintained—in particular, that it can be borne by a militant citizenship. If every citizen becomes a fighter for the cause, in this patriotic, rally-to-the-flag atmosphere, he argues, it becomes very difficult to wage a frontal opposition. But it is not watertight either: if the militant super-citizens find out that the 'measures' are in fact harming their health, then something else will happen. Hurricane Katrina in New Orleans was very illustrative in that regard, because even the 'heroes' (police and emergency workers) sometimes did not show up because their own families were in danger. More generally, people will accept, albeit grudgingly, the lockdowns and measures such as face masks, social distancing, limitations of visits, and curfew, all equally absurd, but draw the line at 'vaccination,' when one's own body has to be entrusted to the authorities. In fact, even this possible public red line has turned out to be largely imaginary.

THE ROCKEFELLER FOUNDATION'S LOCK STEP SCENARIO

Zylberman's book was finally published in French in 2013 but, as mentioned, its conclusions circulated in the form of papers in 2009–10 in the networks of which the Munk institutions at the University of Toronto form a hub. In the same period, in 2010, his idea of scenarios obtained a spectacular implementation with the joint report of the Rockefeller Foundation and the Global Business Network (GBN), a subsidiary of Deloitte involved in scenario development. Spectacular, also because for us as witnesses to the Covid disaster, it is almost unbelievable to read in great detail in a document compiled ten years ago the scenario for what is happening now.

In her preface to the report the president of the Rockefeller Foundation praises scenario planning as the best way to transcend status quo thinking by providing narratives that concern the future. Groups and individuals must be guided in this creative process, and with the help of a story (that is, a narrative with a certain internal logic), events can be set in motion that give current developments a desired twist, often in surprising ways.[69] (We will later meet the author of the foreword on behalf of GBN, Peter Schwartz, again as director of strategy for Salesforce, an IT company that has entered the market for vaccine passports and biometric identification).

We have already seen that after the collapse of the Soviet contender bloc and communism, fear scenarios became a substitute for the concepts of control by which power had been exercised in the preceding period. Instead of a more or less stable constellation of classes whose (unwritten) social contract is expressed in a quasi-governmental program that finds broad acceptance, worst case scenarios evoke visions of fear to which the citizenry submits in the belief that there is no way of escaping the looming disaster it is told hangs over their head, other than by obeying government instructions. Whereas a concept of control derives its logic from the actual relationships of force between the classes, and thus loses effectiveness once they change fundamentally, a scenario relies on the logic of its own storyline. Its authors can only hope that it will also come across as 'believable,' although they need not worry about it being undermined by the media, which after all forms a central part of the fraction at the heart of the new power bloc. But as Zylberman has outlined so eloquently, we are now in the realm of fantasy, the storyline is essentially made up. It is therefore important, as we will see in Chapter 6, that Edelman, the public relations firm that established its reputation for developing such narratives as an advertising tool, played such a defining role in preparing the current Covid crisis. In early 2021, the same agency came to the conclusion, based on surveys in 28 countries, that the corona scenario in the end was a failure. We will return to that later.

The Rockefeller/GBN report discusses four scenarios, three of which don't really matter. Scenario 1 is about a world in which everything is running smoothly and everyone is happily working together, while 3 and 4 each describe a complete collapse, one economic and the other political, so no storyline will be of much help there. The scenario that should concern us here is No. 2, Lock Step. It describes a 'world of tighter top-down government control and more authoritarian leadership, with limited innovation and growing citizen pushback.'

The premise is an extremely virulent influenza pandemic that started in China and was set to strike in 2012. Almost 20 percent of the world's population becomes infected with 8 million dead—in short, a true disaster. The United States (I follow the report) would handle the emergency very poorly, with all kinds of half measures; the Chinese government, on the other hand, responded swiftly. It effectively enforced 'mandatory quarantine for all civilians, as well as the instant and near hermetic sealing off of all borders.' This 'saved millions of lives, stopping the spread of the virus far earlier than in other countries and enabling a swifter post pandemic recovery.'

Meanwhile, still according to the Rockefeller/GBN report, more and more governments had taken measures such as face masks in shops and public areas; and even after the pandemic had subsided, these measures remained in place. In fact, authoritarian surveillance of citizens and their activities was even intensified. All exactly as we are experiencing today. Leaders around the world tightened the reins 'in order to protect themselves from the spread of increasingly global problems—from pandemics and transnational terrorism to environmental crises and rising poverty.'

It is not too difficult to decode the 'increasingly global problems' that leaders around the world had to protect themselves from—growing unrest among the global population, which we identified as the main reason for establishing the Covid state of exception, and the threat of uprisings. What are 'transnational terrorism' and 'increasing poverty' other than keywords to indicate the cause of that unrest and its potential eruption into violence?

Lock Step is the remedy. 'At first,' the report on the fictional lock-downs in 2012 tells us, 'the notion of a more controlled world gained wide acceptance and approval. Citizens willingly gave up some of their sovereignty—and their privacy—to more paternalistic states in exchange for greater safety and stability.'

This echoes Zylberman's conclusions: the majority *want* the lockdown. 'Citizens were tolerant, and even eager, for top-down direction and oversight, and national leaders had often unexpected latitude to impose order in the ways they saw fit. In developed countries, this heightened oversight took many forms: biometric IDs for all citizens were among them.'[70] Certainly the report acknowledged that the population would not accept these restrictions indefinitely. It predicted, however, that since people would not really come to reject the authoritarian turn until 2025 (13 years after the fictitious disaster year of 2012), the changes would have become irreversible in the meantime and the new regime would be so firmly entrenched that a return to the preceding normal would no longer be possible.

The Lock Step scenario also identifies the IT revolution as the basis on which intensive surveillance becomes possible, and by which the existing order can be made secure. To overcome the crisis, the report therefore proposes to transform the economy by basing it on the achievements of the IT revolution entirely, for in this domain, the West still holds all the cards, thanks to the Internet monopolies with their large research budgets. Today we recognize this project as *The Great Reset*, propagated by the World Economic Forum—which it has been advocating ever since the 1990s, under different names but with the same content (a digital world economy). Thanks to the fact that the IT sector is primarily based in the West, major steps forward can be taken here. Countries like Russia and India, on the other hand, will only be able to protect themselves against this Western superiority by protectionist measures, the Rockefeller/GBN report claims. Note that China is not included in this scenario as an antagonist in the same way, because at the time the report was published, relations with Beijing were still relatively good (as indicated also in the compliments for China for its decisive action against the fictitious pandemic).

This Rockefeller/GBN report alone would be enough to show that the measures taken in 2020 were in print long before Covid-19 hit, and that they were seen as the answer to the challenges the West faced when the 2008's financial collapse triggered the global unrest I see as primary issue confronting the capitalist regime, rather than the alleged medical emergency.

The European Union all along has been an important player in these developments, for the social unrest was greatest there, especially in France, whilst the necessary conditions to develop a productive response to it had been largely exhausted, certainly in southern Europe. In 2012 the EU published a remarkable comic book, *Infected*, about the spread of a dangerous virus from China, there presented as an adventure story. In the postscript to this comic book, very unusual for an EU publication to say the least, it became clear that this was not just intended as entertainment (why would the EU publish such?), but as propaganda. It mentions the International Ministerial Conference on bird flu in Beijing in January 2006, where more than 100 countries signed a statement committing themselves to developing and sharing information about national plans against the epidemic. This was in line with the WHO Health Regulations of 2005, a year earlier. In addition, under the heading 'communication,' the postscript states that it might be better if experts in these matters admitted that they do not know everything exactly. This would be the best strategy, according to the EU, to gain public confidence. Of course, that doesn't make it any easier for governments, but the alternative is to lose public confidence due to the likelihood of ensuing conflicting statements.[71] We now know that the government and the experts can declare one month that face masks are unnecessary and even harmful, and the other month the opposite, and that while there may be public confusion, there are hardly any complaints indicating distrust of authorities' motives.

The Rockefeller GBN Lock Step scenario in 2010, Zylberman's reports during the preceding period, and the development of experimental gene therapies that would later be sold to us as 'vaccines,' even the EU comic book about the virus infection coming from China, all indicated that for the Western ruling class, a biopolitical state of emergency had become a serious option. Whether China would be an ally or an antagonist in this undertaking, however, remained to be seen.

Endnotes

1. Robert Miller, 'The Coronavirus is Man-Made According to Luc Montagnier, the Man Who Discovered HIV,' *Gilmore Health News*, 16 April 2020 (online); Pierre Lescaudron, 'Compelling Evidence That SARS-CoV-2 Was Man-Made,' *Sott.net*, 26 June 2020 (online).

2. Frank Ryan, *Virolution* (London: HarperCollins, 2009).

3. J.D. Bernal, *Science in History* [4 vols.] *Vol. 3: The Natural Sciences in Our Time* (Harmondsworth: Penguin, 1969 [1954]), p. 918.

4. Judy Mikovits and Kent Heckenlively, *Plague of Corruption: Restoring Faith in the Promise of Science* [foreword, Robert F. Kennedy, Jr.] (New York: Skyhorse, for Children's Health Defense, 2020), pp. 199–201; Ryan, *Virolution*, pp. 75, 118.

5. Ryan, *Virolution*, pp. 81, 129; Lescaudron, 'Compelling Evidence That SARS-CoV-2 Was Man-Made.'

6. Ryan, *Virolution*, pp. 133–34.

7. Michael Friedman, 'Metabolic Rift and Human Microbiome,' *Monthly Review*, 70 (3), 2018, p. 75; Ryan, *Virolution*, pp. 177–79.

8. Ryan, *Virolution*, p. 27.

9. Ibid., pp. 69–70, 18, 88–89.

10. Jared Diamond, *Guns, Germs and Steel: A Short History of Everybody For the Last 13,000 Years* (New York: Vintage, 1998), pp. 77–78.

11. Mikovits and Heckenlively, *Plague of Corruption*, pp. 141 e.v.

12. Annie Riley Hale, *The Medical Voodoo* (New York: Gotham House, 1936), p. 179.

13. Patrick Zylberman, *Tempêtes microbiennes: Essai sur la politique de sécurité sanitaire dans le monde transatlantique* (Paris: Gallimard, 2013), p. 188.

14. *Reuters Fact-check*: 'Vaccines Caused 1918 Influenza' n.d. (online).

15. David Klooz, *The Covid-19 Conundrum* (n.p. [Apple Books], 2020), pp. 6–9.

16. Peter C. Gøtzsche, 'Our prescription drugs kill us in large numbers,' *PubMed.gov*, 30 Ocktober 2014 (online).

17. Ilana Löwy and Patrick Zylberman, 'Introduction: Medicine as a Social Instrument: Rockefeller Foundation, 1913–45,' *Studies in History and Philosophy of Science*, 31 (3, special issue), 2000, pp. 367–68, 371.

18. Klooz, *The Covid-19 Conundrum*, pp. 10–11.

19. Gary G. Kohls, 'Did psychopath Rockefeller create the Spanish Flue pandemic of 1918?,' *Fort Russ News*, 22 May 2020 (online); Löwy and Zylberman, 'Introduction: Medicine as a Social Instrument,' p. 377.

20. Klooz, *The Covid-19 Conundrum*, p. 26; Kohls, 'Did psychopath Rockefeller create the Spanish Flue?'; Sal Martingano, 'The 1918 "Spanish Flu": Only The Vaccinated Died,' n.d. (online).

21. Klooz, *The Covid-19 Conundrum*, pp. 25–26; Hale, *The Medical Voodoo*, p. 179.

22. Zylberman, *Tempêtes microbiennes*, p. 49.

23. Ibid., pp. 29, 71, 489.

24. Walter Van Rossum, *Meine Pandemie mit Professor Drosten: Vom Tod der Aufklärung unter Laborbedingungen* (Neuenkirchen: Rubikon, 2021), pp. 107–11; Zylberman, *Tempêtes microbiennes*, pp. 129–32.

25. Zylberman, *Tempêtes microbiennes*, p. 187; Catherine Austin Fitts, 'The Injection Fraud—It's Not a Vaccine,' *Global Research*, 1 January 2021 [orig. 28 May 2020] (online).

26. Whitney Webb, 'All Roads Lead to Dark Winter,' *Unlimited Hangout*, 1 April 2020 (online); Zylberman, *Tempêtes microbiennes*, pp. 129–32.

27. Zylberman, *Tempêtes microbiennes*, p. 92 ; Webb, 'All Roads Lead to Dark Winter.'

28. *Medical News Today*, 'Anthrax Vaccine For Soldiers Serving In Iraq, Afghanistan And South Korea To Resume,' 17 October 2006 (online).

29. Webb, 'All Roads Lead to Dark Winter'; Zylberman, *Tempêtes microbiennes*, p. 439.

30. Zylberman, *Tempêtes microbiennes*, p. 405.

31. Ibid., pp. 224, 434.

32. F. William Engdahl, 'WHO "Swine Flu Pope" under investigation for gross conflict of interest,' *Voltaire Network*, 9 December 2009 (online).

33. 'Accomplished pharma prof thrown in psych hospital after questioning official Covid narrative,' *LifeSite*, 11 December 2020 (online).

34. Sonia Shah, 'D'oú viennent les coronavirus? Contre les pandémies, l'écologie,' *Le Monde Diplomatique*, March 2020 [orig. in *The Nation*].

35. Mikovits and Heckenlively, *Plague of Corruption*, pp. 141–61; Ryan, *Virolution*, p. 77.

36. Zylberman, *Tempêtes microbiennes*, pp. 167–68, 171, 181.

37. See my *Global Rivalries from the Cold War to Iraq* (London: Pluto Press, 2006, chapters 4 and 5.

38. James A. Paul, 'Der Weg zum *Global Compact*: Zur Annäherung von UNO und multinationalen Unternehmen' [trans. Th. Siebold], in Tanja Brühl *et al.*, eds., *Die Privatisierung der Weltpolitik: Entstaatlichung und Kommerzialisierung im Globalisierungsprozess* (Bonn: Dietz, 2001), pp. 114–15.

39. Ibid., pp. 126–27.

40. Daan de Wit, *Dossier Mexicaanse griep: Een kleine griep met grote gevolgen* (Rotterdam: Lemniscaat, 2010), p. 113.

41. Paul, 'Der Weg zum *Global Compact*,' pp. 107–8.

42. Rosemary Frei, 'Did Bill Gates Just Reveal the Reason Behind the Lock-Downs?,' *OffGuardian*, 4 April 2020 (online); De Wit, *Dossier Mexicaanse griep*, pp. 98–99.

43. Joseph A. Camilleri and Jim Falk, *Worlds in Transition: Evolving Governance Across a Stressed Planet* (Cheltenham: Edward Elgar, 2009), pp. 413–14.

44. Zylberman, *Tempêtes microbiennes*, p. 183.

45. See my 'The Sovereignty of Capital Impaired: Social Forces and Codes of Conduct for Multinational Corporations,' in Henk Overbeek, ed., *Restructuring Hegemony in the Global Political Economy: The Rise of Transnational Neo-Liberalism in the 1980s* (London: Routledge, 1993).

46. De Wit, *Dossier Mexicaanse griep*, p. 95; Zylberman, *Tempêtes microbiennes*, pp. 37, 226.

47. Camilleri and Falk, *Worlds in Transition*, pp. 396–97; De Wit, *Dossier Mexicaanse griep*, p. 12.

48. Zylberman, *Tempêtes microbiennes*, pp. 72, 213.

49. De Wit, *Dossier Mexicaanse griep*, pp. 96–97, 126–27; Engdahl, 'WHO "Swine Flu Pope" under investigation.'

50. Camilleri and Falk, *Worlds in Transition*. pp. 404–5.

51. *WHO, International Health Regulations,* 3rd ed. (Geneva: World Health Organization, 2005 [1969]), p. 1; Camilleri and Falk, *Worlds in Transition*, pp. 415–16.

52. *WHO, International Health Regulations*, p. 17.

53. Camilleri and Falk, *Worlds in Transition*, p. 418.

54. Zylberman, *Tempêtes microbiennes*, pp. 484–85.

55. Camilleri and Falk, *Worlds in Transition*, p. 424.

56. Van Rossum, *Meine Pandemie mit Professor Drosten*, pp. 67, 75–77.

57. Engdahl, 'WHO 'Swine Flu Pope' under investigation'; De Wit, *Dossier Mexicaanse griep*, pp. 20–21.

58. Van Rossum, *Meine Pandemie mit Professor Drosten*, pp. 230–31.

59. De Wit, *Dossier Mexicaanse griep*, pp. 25–27, 29.

60. Ibid., pp. 57, 64–65, 82–84; Van Rossum, *Meine Pandemie mit Professor Drosten*, pp. 78–79.

61. De Wit, *Dossier Mexicaanse griep*, pp. 143, 200, 201–3; Flo Osrainik, *Das Corona-Dossier: Unter falscher Flagge gegen Freiheit, Menschenrechte und Demokratie* [voorwoord Ullrich Mies] (Neuenkirchen: Rubikon, 2021), pp. 178–80.

62. Ulrich Beck, *Risikogesellschaft: Auf dem Weg in eine andere Moderne* (Frankfurt: Suhrkamp, 1986).

63. Peter Phillips, *Giants: The Global Power Elite* [foreword, W.I. Robinson] (New York: Seven Stories Press, 2018), p. 124.

64. See my *The Discipline of Western Supremacy*, Vol. III of *Modes of Foreign Relations and Political Economy* (London: Pluto Press, 2014), pp. 229–30.

65. Zylberman, *Tempêtes microbiennes*, p. 152.

66. Ibid., pp. 425–27.

67. Ibid., pp. 415, 402–3.

68. Ibid., pp. 430–31.

69. *Rockefeller Foundation* and *Global Business Network, Scenarios for the Future of Technology and International Development* [Judith Rodin, Peter Schwartz, forewords] (New York: Rockefeller Foundation and San Francisco: Global Business Network, 2010), p. 9.

70. Ibid., pp. 18–19.

71. EU Publications Office, *Infected* [pdf], 31 January 2012 (online).

5.

BIOLOGICAL WARFARE WITH OR AGAINST CHINA?

In addition to the prospect of a revolt of the peoples after 2008–10, threatening the ruling classes around the world, as well as the possibility of a new financial collapse, a further world-historical challenge has emerged of specific concern to the ruling class in the West: the rise of China.

After Britain had opened the millennium-old Empire to imperialist plunder with the Opium Wars in the mid-19th century, China managed to restore its independence via the 1949 revolution. Led by the communist party under Mao Zedong and Zhou En-lai and initially as an ally of the Soviet Union, China too developed as a contender state. The lack of substantial democracy proved to be the Achilles' heel for both types of state socialism because it also squeezed out the peoples' initiative and creativity, without which a society is eventually condemned to stagnation. In the late 1970s the Chinese state leadership solved this by initiating a partial capitalist restoration through a revolution from above, a passive revolution, the characteristic mode of change in a contender state. The state class led by Deng Xiaoping then made China's labor reserve available for capital without relinquishing state power entirely. The suppression of the 1989 student uprising, memorialized by the image of the tanks in Tiananmen square, pre-vented the state from losing control completely.[1]

In this chapter, we first briefly discuss Chinese aspirations in the IT domain and how these are being responded to by the West. I then turn to how the United States has developed a capacity for biological warfare, in which, as in the international division of labor between the U.S. and China, cooperation and rivalry are in-tertwined. While it is quite possible that SARS-2 was the result of a U.S. project subcontracting dangerous virus research to China, there are other explanations that need to be taken into account as well.

THE CHINESE CHALLENGE

In October 2016, journalist William Engdahl wrote an optimistic reflection on China's plans for a New Silk Road, or One Belt, One Road (OBOR) project, presented three years earlier by President Xi Jinping on a visit to Kazakhstan.[2] Now underway, this project entails the creation of a network of high-speed rail lines that will traverse the vast landmass of Eurasia, from the coast of China and Russia, through Central Asia, and on to northwestern Europe, with Rotterdam as one of the final destinations. In addition, it includes shipping routes between China via new super-ports to Piraeus, the port of Athens taken over by China when Greece went on sale on the instruction of the European Union and the IMF, and elsewhere. The costs for these gigantic projects will be borne by the major Chinese banks, the Asian Infrastructure Investment Bank (AIIB) created by China for the purpose, and other sources.

With this initiative, Engdahl wrote, the doctrine of the British geographer, Halford Mackinder, that the Eurasian landmass should never be allowed to coalesce into a single, political-eco-nomic entity, can be buried. Anglo-American world strategy has been based on that doctrine for more than a century, but neither the bankrupt United States nor the near-bankrupt European Union today are in a position to uphold that goal any longer. Europe of course has always served as the lever against Eurasian unity, in particular by preventing a rapprochement between Germany and Russia, which earlier culminated in the two world wars.

China's advance must therefore be halted, but where the relationship with Putin's Russia is unambiguous in this respect (i.e., undermining Russian plans for a Eurasian Union and even undisguised plans to overthrow the president himself, etc.),[3] the relationship with China is ambivalent. On the one hand, China is the 'workshop of the world,' where large-scale investments are made by Western companies such as Apple and others because of cheap Chinese labor. This has worked to buy social peace in the West, as incomes stagnate there, insofar as a cheaper iPad or washing machine, clothing and sports shoes produced with Chinese wages make the loss of Western purchasing power less dire. The Chinese state, in turn, has been supporting U.S. deficits by investing in Treasury Bonds, although China, like many other non-Western states, is increasingly reluctant to continue to support the dollar in this way. Barter agreements with Russia, the use of their own currencies, and a shift to gold as a monetary reserve, are all signs of the flight out of the dollar without a clear alternative yet taking shape.

The financial collapse of 2008 also marked the moment when the Chinese challenge became acute and the dilemma of choosing between ongoing synchronization or open rivalry could no longer be evaded. As David Lane shows, by 2007 the share of world GDP of the BRICS bloc (Brazil, Russia, India, China and South Africa) had overtaken the shares of NAFTA (U.S., Canada and Mexico) and of the EU; China's GDP alone overtook that of the U.S. in 2014. After 2008, China's hi-tech exports overtook those of the United States.[4] China has also taken important steps in the microbiological field, such as the use of bacteria as a storage medium for information, thus allowing it to make a giant leap forward in the field of quantum computing.

Advice on how the West should approach China has been forthcoming from all angles, but with a Democratic president now in the White House, the view of the party's longtime geopolitical ideologue, Zbigniew Brzezinski, mentioned several times already, is of special significance. His last book, the aforementioned *Strategic Vision: America and the Crisis of Global Power,* was published in 2012 (he passed away in 2017). In this book Brzezinski argues that the West (that is, the U.S., the English-speaking rich countries, and the EU) is locked in a historic crisis. Indeed, he saw many similarities between the America of the first decades of the 21st century and the Soviet Union on the eve of its collapse. Yet Brzezinski simultaneously indicated that it would depend on the United States whether a turnaround can be expected in the decline of the West, a downturn he saw taking place in a number of closely intertwined areas. The idea that China could take over the global role of the U.S. is an illusion, according to Brzezinski. Even with continued progress and increasing influence in the world, there will be no *Pax Sinica* to succeed the malfunctioning *Pax Americana*, only chaos. This, along with the unrest among the peoples, excessive speculation practices, and the rise in global temperatures, will contribute to extreme instability.[5]

In the IT sector, the Chinese advance is being fought openly. China has taken a lead in some areas (notably by the introduction of the super-fast 5G network, which makes surveillance of the population through facial recognition and other means even more efficient, albeit with enormous risks to public health).[6] Since May 2019, Washington has identified the top Chinese companies in this field, Huawei and ZTE, as a threat to national security; they are no longer allowed to operate in the U.S. Subsequently, production in China itself came into focus. While the 5G base stations that Huawei exports do not use U.S. technology, in August 2019 Washington banned the supply of microchips to Huawei, cutting the company off from the required parts not only for its 5G equipment, but also for its smart phones. This affects suppliers such as the largest microchip maker in the world, Taiwan Semiconductor Manufacturing Co. (TSMC), as well as Samsung (South Korea) and Sony (Japan), which supplies screen sensors to Huawei. The microchip producers such as TSMC, Intel and Samsung are in turn dependent on Japanese companies such as Nikon and Dutch

ASML who supply the machines with which circuits are printed on silicon discs (steppers).[7]

Reorganizing the IT product chain was thus meant to isolate China, and Washington hoped to get the West on board, without shunning any means. In December 2018, at U.S. request, Canada kidnapped the daughter of Huawei's founder, who also was the company's financial director, for alleged violation of the U.S. embargo against Iran. Great Britain, which had given Huawei permission to become active in the British 5G network in January 2020, reversed this under pressure from Washington. The Netherlands, which with ASML is an important link in the global IT product chain, is under great pressure from the U.S. to keep Huawei away from its projected 5G network. By making a distinction between the 'critical core' of the network and peripheral facilities, the Dutch government hopes to avoid millions of losses on contracts. The European Commission in turn is also trying to limit the damage in this way, with this instruction to governments to not allow politically unreliable (read: Chinese) companies into the critical core and go for Ericsson or Nokia instead.[8]

AMERICAN BIOLOGICAL WARFARE RESEARCH

While the U.S. and the West in the evolving IT field openly respond to Chinese IT development with economic warfare (the U.S. more so than the rest of the West), there is a much more ambivalent relationship in the field of microbiological research. This even concerns the extremely sensitive area of bioweapons research.

Biological research into viruses and bacteria serves many purposes, from the search for vaccines and medications to the development of pathogens for use on military or enemy civilian targets. In an interview in early 2020, Francis Boyle, a professor of international law at the University of Illinois and author of the U.S. legislation implementing the 1972 Biological Weapons Convention, mentioned the sum of $100 billion the United States has spent on biological weapons after 9/11 (until 2015). In constant dollars, that would be roughly equivalent to the cost of the Manhattan Project to develop the first nuclear weapons. According to Boyle, there are about 13,000 biologists working in defense-related research in the U.S.[9] Such research is not easily classified as either offensive or defensive, since the same sort of work can serve both purposes. Both the development of weapons and the discovery of vaccines or treatments are of military significance, since treatments are required to protect the domestic population in order to enable their military use.

The Americans have had their own biological warfare laboratory at Fort Detrick, Maryland, since 1942. Experiments were conducted there related to anthrax, botulism, plague, tularemia, Q fever, and other diseases. During the war, Japan in particular had deployed biological weapons on a large scale. It had signed, but not ratified, the 1925 Geneva Protocol banning biological weapons; in World War II, Unit 731 of the Japanese Imperial Army conducted deadly experiments with cholera and other diseases, and dropped so-called 'plague bombs' on Chinese cities.[10] After the war, Japanese army biochemists were granted immunity from prosecution if they transferred their knowledge to Fort Detrick; the same arrangement was offered via Operation Paperclip to Erich Traub, the chief of the German bio-weapons laboratory on the island of Riems, who worked on foot-and-mouth disease and Newcastle disease among others. Traub, who also worked for the Russians immediately after the war, provided the Americans with information that formed the basis for animal testing on Plum Island, Fort Detrick's domestic offshore biowarfare testing ground.[11] This enabled the American army to bombard North Korea with bacteriological weapons during the Korean War (1950–54). Insects and voles infected with pathogens (bubonic plague and others) were thrown from airplanes in nighttime bombings.[12]

In 1969, President Richard Nixon ordered the destruction of all U.S. biological weapons. He declared that the United States would refrain from using these weapons and that biological research would be conducted only for protection, such as immunization and other security measures. The ratification of the 1972 Biological Weapons Convention should have ended the biowarfare programs of the major signatories (including the USSR and Great Britain) once and for all, but in 2001 it was revealed that the U.S. had continued to work on biological weapons.[13]

To make the threat posed by biological weapons credible, the CIA had funneled false intelligence about mysterious, top-secret U.S. chemical and biological warfare programs to Moscow. Thus, it was hoped to entice the Soviet Union to waste its increasingly scarce resources on pointless projects, and the Soviet leadership indeed started its own bio-military program in 1973. The best biochemical specialists were deployed under the guise of pharmaceutical research (the 'BioPreparat' program). This in turn provided the justification for the re-launching of the U.S. bioweapon program as defectors were brought before Congress to reveal what was happening in the USSR (the same thing happened again after the collapse of the Soviet Union). Impressed by these reports, the Reagan administration, and later Clinton, decided to hold on to the supplies of the smallpox virus in a lab in Atlanta, Georgia, among other things. By the time George W. Bush scrapped the biological weapons verification protocol in July 2001, the 1972 Biological Weapons Convention was effectively canceled.[14]

In response to questions in Congress, it emerged little by little (especially in the 'critical' 70s in the aftermath of Watergate when the Senate Committee led by Frank Church exposed military and intelligence agency interference with regular democratic procedure and civil rights) which pathogens had been worked on in the past period, including experiments with Lyme disease, which is still afflicting Americans. The overarching program was MKULTRA, already mentioned in Chapter 1. It investigated, among other things, the use of psycho-pharmaceuticals for use in interrogation. The biowarfare program at Fort Detrick was called MKNAOMI, a shared responsibility of the Department of Defense and the CIA.[15] The aforementioned CIA chemist, Sydney Gottlieb, had been a PhD student at the University of Wisconsin of Ira Baldwin, who later became research director at

Fort Detrick. However, MKNAOMI remains shrouded in mystery because Richard Helms, the director of the CIA fired by Nixon in 1973, ordered all documentation surrounding the program to be destroyed.[16] In the same year Gottlieb resigned from the CIA, making it less likely that he was still involved in the use of viruses in the CIA's illegal war in Zaire and Angola, as has been alleged.[17] In contrast, the Apartheid regime in South Africa, according to a 1998 document, did play a role in this with the South African Institute for Maritime Research (SAIMR); after the fall of Apartheid, SAIMR's biological warfare research results were handed over to Fort Detrick.[18]

Gottlieb was also involved in studies of the use of drugs to manipulate the human mind, such as LSD, the use of lie detectors, etc. As part of MKULTRA, he conducted experiments to which the Allan Memorial Institute in Montreal also contributed. The supposed termination of MKULTRA after disclosures in 1964 was later discovered to be only its name change to MKSEARCH. Another branch of MKULTRA, ARTICHOKE, aimed to put a person in a state where he would follow instructions against his will. In the murder of Robert Kennedy, this may have been the case with the perpetrator, Sirhan Sirhan.[19] The defense research agency, DARPA, has brought research into the brainwashing techniques of the Korean War era and the biological research into manipulating brain functions together again.[20] Since the SARS-2 laboratory virus also has psychotropic effects (from loss of smell and taste to panic attacks), this is a possible indication that Covid-19 is also associated with this type of biological warfare. We will come back to this later.

BIOWEAPONS AND THE WAR ON TERROR

In early October 2001, in the aftermath of the 9/11 attacks, letters containing anthrax were delivered to the Senate Democrats' leader Tom Daschle and his colleague, Patrick Leahy (and to some media). Anthrax was a major field of biological warfare research, and under Clinton, but without his knowledge, the U.S. military was working on vaccine-resistant variants. When the president found out, he ordered this to be stopped, but the project was resumed under the Bush administration.[21]

Fort Detrick, meanwhile, had been renamed U.S. Army Medical Research Institute of Infectious Diseases (USAMRIID), and its biowarfare programs continued under the cover of the policy of non-proliferation, i.e. to prevent the spread of bioweapons. USAMRIID maintained stores of toxins and cultures of anthrax, tularemia, brucellosis, Venezuelan equine encephalitis. and small-pox.[22] In fact, Fort Detrick had remained a disposal site for biological and chemical weapons because it was the best repository for those weapons of war and the military had even used the area around it for training, e.g. for Agent Orange, which was used in Vietnam.[23] In a confidential NSA newsletter of 6 November 2003, the year of the invasion of Iraq, USAMRIID's work is referred to as the 'U.S. Army biological weapons research facility' working with the Armed Forces Medical Intelligence Center (AFMIC) of the Defense Intelligence Agency (DIA).

The anthrax letters to Daschle and Leahy were attributed by Israeli and U.S. experts to Saddam Hussein of Iraq, but Professor Francis Boyle, quoted earlier, made it known in the media that the anthrax was most likely from a U.S. bio-weapons laboratory,[24] which led to his being blacklisted by the media. But the anthrax variant in question (the Ames type) had indeed been developed by the U.S., and the denial that it was a home-grown substance persisted because it obviously raised doubt about the 9/11 attacks themselves.[25] The FBI went on to identify a lab technician, Bruce Ivins, who worked at USAMRIID, as the perpetrator but the evidence was questionable. After it was brought to his attention that he would be accused of terrorism, Ivins took his own life in 2008.[26]

Meanwhile, within a month of 9/11, an entirely new branch of defense industry, based not on fighter jets but on genetics, was born. Three years later a private bio-security sector had also constituted itself, with annual sales of $75 billion by 2010. Venture capital invested in biotech start-ups was quick to capitalize on the U.S. defense effort in microbiology, which later became the BioShield program. In total, about 10 billion had been invested in bio-protection by that time; a company like VaxGen saw its share price rise by 20 percent after it secured a government contract of nearly a billion for anthrax vaccines, to name but one example. Rumsfeld's 'revolution in military affairs,' initiated under George W. Bush, also brought about a 'revolution in public health' by its impetus to research.[27]

Although the U.S. had withdrawn from the negotiation process for verification of biological weapons of mass destruction, Washington did support UN Security Council Resolution 1540 in 2004, which tightened the control regime over weapons of mass destruction, including biological weapons. However, that resolution specifically targeted 'rogue states' harboring terrorists; the U.S. itself was not subject to inspection although it continued to work in this area. A 2010 U.S. Air Force report speculated on the threat of 'binary biological weapons, designer genes, gene therapy as a weapon, stealth viruses, host-swapping diseases and designer diseases,' indicating continued interest in these types of weapons.[28] The global tracking of all telephone and Internet traffic disclosed by Edward Snowden in 2013 also extends to infectious diseases. The NSA eavesdrops on all communications from the WHO, Doctors Without Borders and the International Red Cross and is notified of any disease outbreak without delay. This was the case with the SARS-1 epidemic in China, cholera in Liberia and a series of epidemics in Iraq. The NSA's Target Office of Primary Interest (TOPI) has the ministries of health, hospitals, international and local Red Cross and Red Crescent departments in affected countries under continuous surveillance.[29] This is also important because both the Red Cross and Red Crescent are represented at the highest level on the Global Preparedness Monitoring Board, the WHO and World Bank body that warned the world in 2019 that in the event of a pandemic, the commitments entered into in 2005 would come into effect, as we will see in the following chapter.

In addition to research done at Fort Detrick, the United States Army produces and tests biologically active compounds for military use on a dedicated base, the Dugway Proving Ground, at the West Desert Test Center in the state of Utah. That base is supervised by the Army Test and Evaluation Command. The Life Sciences Division of Dugway Proving Ground is responsible for the production of bio-agents; these are again being tested for use as aerosols at the Lothar Saloman Life Sciences Test Facility. This includes an aerosol technology department and a microbiology department. The Aerosol Technology Branch develops aerosols with biologically active substances, i.e. toxins, bacteria, viruses and related organisms, for use in testing. Since aerosols are a mode of delivery, this cannot be characterized as 'protection.'[30]

One of the companies working at the National Biodefense Analysis and Countermeasures Center (NBACC) in Fort Detrick, is Battelle Memorial Institute, a private company also active in post-Soviet Georgia (see below). In Fort Detrick it operates under a contract from the Department of Homeland Security awarded for 2006–16 and a smaller contract for 2015–26. The experiments at Fort Detrick include testing of toxins in aerosols, powder dissemination and testing meliodosis, a viral disease with the potential of a biological weapon, on primates. Battelle was already producing other substances at bio-safety level 4 (the highest, P4).

DARPA's most recent program, the Insect Allies program, is also not a project for defensive purposes; it aims to develop a 'new class of biological weapons.' This was the conclusion of a group of scientists led by Richard Guy Reeves, of the Max Planck Institute for Evolutionary Biology in Germany, in an article in *Science*. In that article they warned that the use of insects as a vehicle for horizontal genetic means of modification in the environment (HEGAAS) made it clear that the intention was to use genetic modification as a weapon in this way.[31] Transgenic manipulation of wasps, bees and mosquitoes could be developed to transmit protein-based biological agents on a large scale.[32]

The United States Army Chemical Research and Development Command, Biological Weapons Branch, studied biting mosquitoes in a number of field tests at the Dugway Proving Ground in the early 1960s. USAMRIID experimented as late as 1982 with sand flies and mosquitoes that could be carriers of Rift Valley virus, dengue, Chikungunya and equine encephalitis, all viruses that the U.S. military have investigated for their potential as biological weapons. Bulgarian journalist Dilyana Gaytandzhieva writes that in a 1981 report, the U.S. military compared two scenarios: simultaneous attacks on a city by *Aedes Aegypti* mosquitoes infected with Yellow Fever and an attack with tularemia aerosols. The effectiveness of these studies were then compared in terms of costs and numbers of victims. The Zika virus, which causes genetic malformations in newborns and recently emerged in Latin America, is one of the diseases transmitted by *Aedes Aegypti*, also known as the Yellow Fever mosquito. In 2003, during the invasion of Iraq, American soldiers were bitten by sand flies and contracted Leishmoniasis, the acute form of which, if left untreated, can be fatal.[33]

It would be interesting to know how often these weapons of mass destruction, the 'B' of ABC weapons, have come up in debate concerning defense budgets in any parliament, beginning with the U.S. Congress itself. Certainly, the claim that SARS-2 came from a laboratory becomes a lot less exotic when we consider the existence of these extensive research programs, bearing in mind this is just what is being researched in the U.S., stateside.

U.S. OVERSEAS BIOLOGICAL WEAPONS LABS

In addition to such research facilities as Fort Detrick at home, the United States also manages a series of laboratories abroad. These are mainly located on the Russian and Chinese borders, and also in sub-Saharan Africa. Dilyana Gaytandzhieva has made her name mapping these foreign networks in great detail. Although Article 8 of the Rome Statute, by which the International Criminal Court was established, classifies biological experiments for military purposes as a war crime and as we saw, the UN Convention on the elimination of biological weapons dates back to 1972, the Pentagon operates biowarfare laboratories in 25 countries. These laboratories are funded by the Defense Threat Reduction Agency (DTRA) under the Cooperative Biological Engagement Program (CBEP), with a budget of $2.1 billion per year.

Pentagon Biolaboratories in 25 countries around Russia, China, Iran, etc. under DTRA Cooperative Biological Engagement Program (CBEP)
Source: DoD

SOURCE: Dilyana Gaytandzhieva, 'The Pentagon Bio-weapons,' *Dilyana.Bg*, 29 April 2018 (online)

The former Soviet republic of Georgia is important in this regard. The Lugar Center, named after U.S. Senator Richard Lugar, 17 kilometers from the U.S. military air base Vaziani, not far from the capital Tbilisi, is home to biologists from the U.S. Army Medical Research Unit-Georgia (USAMRU-G) as well as private companies. Gaytandzhieva consulted the U.S. federal registry for government contracts and found that research focuses on biological warfare agents such as anthrax, tularemia and viruses such as Crimea-Congo hemorrhagic fever (CCHF), a variant of dengue. In the run-up to the U.S. invasion of Afghanistan CCHF was discovered to be prevalent in both Afghanistan and Pakistan.[34]

Gaytandzhieva identified three private U.S. firms working in the Lugar Center laboratory near the Georgian capital: CH2M Hill, Battelle and Metabiota. Outsourcing research to private companies has the advantage of avoiding congressional scrutiny and eliminating further legal restrictions. Diplomatic status for all researchers (under the U.S.-Georgia defense cooperation agreement of 2002) also grants them immunity from legal liability on the Georgian side. In addition to the Pentagon, these firms also work for the CIA and other U.S. government agencies. CH2M Hill has other DTRA research contracts in Uganda, Tanzania, Iraq, Afghanistan, and Southeast Asia, but the Georgian contract is by far the largest (half the value of all contracts combined).

In 2014, the Lugar Center was equipped with an insect laboratory, which may have something to do with the fact that from 2015, Tbilisi has been plagued by stinging flies, a species that has also emerged in neighboring Dagestan (Russia). As early as 2014, related to another DTRA project, the tropical mosquito *Aedes albopictus* emerged in Georgia as well as in the Krasnodar region of Russia and in Turkey. There is even evidence that tests were conducted in Russia itself: in the spring of 2017, citizens reported that a drone near the Russian-Georgian border was scattering white powder, but neither the Georgian border police nor U.S. officials operating on that border were available for comment.[35]

Biowarfare research is also being conducted for the aforementioned Battelle Memorial Institute, also a subcontractor for the Lugar Center. This company, ranked 23rd of contractors working for the U.S. government, does research and testing for the Pentagon, the CIA, and other government agencies. For the CIA, it is investigating the reuse of small anthrax bombs dating back to Soviet times. Metabiota Inc. also has contracts under the DTRA program in Georgia and Ukraine. The company focuses on global fieldwork, pathogen tracking, outbreaks and clinical trials. During the Ebola crisis in West Africa, it operated under a major contract in Sierra Leone, one of the countries at the heart of the 2012–15 epidemic.

Eleven DTRA-funded bio-laboratories are active in Ukraine, over which the country has no control. This goes back to a 2005 agreement between the Pentagon and Ukraine's Ministry of Health. It prohibits the Kiev government from disclosing sensitive information about the program; Ukraine is required to hand over the dangerous pathogens produced there to the U.S. Department of Defense for further biological research. Finally, the Pentagon has access to Ukraine's state secrets related to the said agreement.[36] One of the Pentagon laboratories is located in Kharkov where at least 20 Ukrainian soldiers succumbed to a flu-like virus within two days in January 2016, and another 200 were hospitalized. However, Kiev did not publicize this incident. Suspicious outbreaks of Hepatitis A in the southeast of the country (where most U.S. laboratories are located) and cases of cholera caused by contaminated drinking water have also been reported.

In 2014 there was an outbreak in Moscow of a new, highly virulent variant of the cholera agent *Vibrio cholera*, related to the species identified in Ukraine. According to a study by a Russian research institute it was caused by identical bacteria, and according to Gaytandzhieva, Southern Research Institute, one of the contractors in Ukraine, works with cholera, influenza and Zika, all pathogens of military interest to the Pentagon. The Southern Research Institute has worked under a DTRA program in Ukraine since 2008 and has also collaborated on a Pentagon program for anthrax, but in that project Advanced Biosystems was in charge. Advanced Biosystems is the company of Ken Alibek, a microbiologist and biowarfare expert from Soviet Kazakhstan who moved to the U.S. after the collapse of the USSR.

The latest DARPA-funded biowarfare research concerns bats as transmitters of deadly human pathogens. Bats are the reservoir of the Ebola virus, SARS, MERS and other viruses that cause deadly diseases, but before they can spread to humans, a number of mutations must occur, more than can occur in nature within a given time frame.[37] According to the *Washington Post,* the Pentagon's interest in the bat research route was prompted by Russian efforts to weaponize bats, but that can safely be put aside. The Soviet Union was indeed engaged in secret research on the Marburg virus, but no bats were involved, and the program ended with the collapse of the USSR.[38]

I have gone into all these labs and research programs so extensively to make it clear that if a virus such as SARS-2 came from a lab, this should be less surprising, given the extensive range of such U.S. biowarfare institutions spread around the world and the nature of their research. The U.S. even undertook extensive collaboration with China in this highly sensitive area.

U.S. COLLABORATION WITH CHINA IN THE MICROBIOLOGICAL FIELD

Because China allied itself with the U.S. against the Soviet Union in the 1970s and allowed a controlled return to capitalist relations, American institutions gained a level of access in China that they did not have in Russia or Iran, for example. This access and the associated cooperation, usually in the nature of subcontracting, also extended to the field of microbiology. This created the paradoxical situation that biological warfare agents, which in principle are also to be directed against China, were partly developed together with it; a situation resembling Beijing's contribution to the financing of the U.S. deficit, helping among other things to make war preparations against China possible.

The extent of this collaboration was illustrated by the 2018 publication of the complete genome of two variants of the coronavirus. This was the result of a project funded by China's Ministry of Science and Technology, the U.S. National Institute of Health, and the development organization USAID (a CIA umbrella organization). One of the conclusions of this study was that existing vaccines against MERS (Middle East Respiratory Syndrome) would have no effect against these viruses. It was therefore proposed, by the Chinese *and* the Americans, to develop new vaccines in advance of their anticipated evolution.

Duke University was another link between the U.S. national security state and China. Duke is a key partner in DARPA's Pandemic Prevention Platform (P3) program and was involved in a study of coronaviruses in bats in Kazakhstan, entirely at the expense of DTRA, i.e. the Pentagon. Here, too, a search was made for viruses related to MERS (the actual MERS epidemic eventually turned out to have caused only 866 deaths between 2012 and

2020, but an outcome of a completely different order can be made possible in the laboratory). In 2018, the university also entered into a cooperation agreement with Wuhan University, and in the same year, Duke Kunshan University in China was established.[39]

The collaboration between USAMRIID in Fort Detrick, Maryland, and the Wuhan Institute for Virology dates back to the 1980s but was intensified after the SARS outbreak in 2003. The Wuhan Institute had a program of experiments on bats that were not bought on the market but had been collected from caves in Yunnan province in South China, nearly 2,000 kilometers south of Hubei (the province of which Wuhan is the capital). The gain-of-function (increase in lethality) study was conducted under the auspices of Dr Anthony Fauci's NIAID and NIH. This also took place in the U.S. itself in a series of research institutions in addition to Fort Detrick, such as the University of North Carolina at Chapel Hill (according to Dr Francis Boyle, research at that institution was conducted into infecting mammals with the spike protein by which coronaviruses attach themselves to host cells to cause respiratory problems).[40] The Erasmus Medical Center in Rotterdam, which is funded by NIAID and NIH, is an important European node in the network in which gain-of-function research is being conducted.[41]

The research consists of artificially inducing mutations in the genome of a virus in order to make it more virulent or more contagious between humans (e.g. by allowing it to spread via aerosols), with a view, it is speciously argued, to giving researchers insight into how such dangerous viruses might evolve in future and providing actual samples to work on to test treatments. Experiments with Enhanced Potential Pandemic Pathogens (PPPs) are legal in the U.S.[42] Fort Detrick uses a new technology, CRISPR-Cas 9, which extracts specific genetic sequences from viruses. This encodes a type of sequence ('cleavage furin site') which, according to some researchers, has also been found in SARS-2 and causes its exceptional virulence.[43]

The Erasmus study, which began in 2010, focused on infecting ferrets, which have a respiratory system functioning like that of humans, with Influenza A virus that had been manipulated to spread through the air, as aerosols. The report on this was published in *Science*, 22 June 2012.[44] In 2011, after he had genetically altered the H5N1 avian influenza strain in this way, Professor Ron Fouchier boasted that he had produced 'probably one of the most dangerous viruses you could make.'[45] Late that same year an Op-Ed piece appeared in the *Washington Post* entitled 'A Flu Virus Risk Worth Taking,' with Fauci as one of the authors, assuring that a vaccine against upgraded viruses is always made at the same time.[46] However, this did not prevent the Obama administration from imposing a moratorium on federal funding for this type of research in October 2014. After repeated serious violations of safety regulations had been established, this moratorium partly met the demand of critics of gain-of-function research to ban it altogether.

Shortly before the moratorium, Fauci had started a research program at the National Institute of Health (NIH) to see if the coronavirus from Chinese bats could be upgraded to spread to a range of other animal species, including 'humanized mice,' mice with a number of human genes added, even human tissue. After the 2014 federal ban had entered into force, Fauci continued the gain-of-function research by privatizing it to the EcoHealth Alliance lab in New York. In 2016, this company had started a multi-billion dollar study, the Global Virome Project, together with USAID, in order 'to be prepared for the next pandemic' and to identify viruses occurring in the wild that might cause it.[47] The director of EcoHealth Alliance, Dr Peter Daszak, had traced the SARS-1 virus to bats and had good contacts in Wuhan. He used $3.7 million (over six years) made available for the NIH project to outsource the gain-of-function portion to the Wuhan Institute of Virology, led by the virologist, Dr Shi Zheng-Li. The program built on previous research for the same grant, bringing the total to $7.4 million.[48]

The outsourcing of the riskiest research to Wuhan was also made possible by a new security level 4 department (P4) in the Institute of Virology that had been built there. This went back to an agreement between France and China in 2004. At the opening of the new facility in 2017, in addition to a French cabinet minister, a top executive of the Institut Pasteur was also present. This institute, which works closely with Sanofi-Pasteur, the vaccine branch of the Sanofi-Aventis pharmaceutical group, which is in the running for an intravenous gene therapy ('vaccine') against Covid, had in fact patented SARS-1 in 2003. According to the French pharmacology professor, Fourtillan, who had contributed to the critical documentary, *Hold-Up*, and had been locked up in a mental institution in December 2020, Institut Pasteur had also patented a SARS-2 in 2011 because the first patent was due to expire in 2023. In both cases, the nominal goal was to develop a vaccine for this upgraded virus. The Institut Pasteur threatened legal action against Fourtillan after he also stated that the French institute was responsible for releasing the virus from the Wuhan laboratory, but although Fourtillan was indeed taken into detention again, it was apparently on account of an unrelated felony.[49] The fact that a patent on (an as yet imaginary?) SARS-CoV-2 had already been filed eight years before its outbreak may explain why all kinds of derivative patents, orders for medication, etc., have been reported on social media with the authentic documentation attached, as noted before.

In the P4 laboratory in Wuhan, several virus variants were isolated from bat droppings, with which human cells were then infected so that the virus could multiply in them. The aim was to find out how the bat virus could be manipulated in such a way that it could attach itself to receptors in human cells (with the spike proteins). For example, SARS-2 is 10 to 20 times stronger in its binding to ACE2 receptors in the human lung than the original bat virus and therefore more easily transmissible between humans. The project in Wuhan produced 13 papers, all co-authored by Shi Zheng-Li, and in 2017 also featuring Daszak as one of the writers. In that article, the project number of Fauci's research is also listed as the source of finance.[50]

In 2017, the federal ban on gain-of-function research was lifted again by the Trump administration. The DARPA research project Preventing Emerging Pathogenic Threats (PREEMPT) was officially announced in April 2018. It focuses on animal reservoirs of diseases, especially in bats, with the justification that it is necessary to remain one step ahead of nature by artificial mutations. By creating super-viruses that are more pathogenic and easier to transmit, the development of these viruses could be better studied; the same for the question how genetic alterations change the way a virus interacts with its host. It also makes it possible to develop antiviral drugs that can prevent a pandemic.

Whether it is really about the defense against pathogens, about bioweapons, about gene therapy as a treatment method ('vaccine') or most likely, about a combination of them all, that is, creating a more virulent pathogen and then developing the most appropriate vaccine, it is certainly an opportunity to make huge profits. After all, the PREEMPT program was concerned with viruses transmitted between humans and with the controversial DNA plasmid and mRNA gene therapy with which these coronavirus types are to be warded off. At least two DARPA studies using this controversial technology are classified and focus on the possible military application of *gene drive* technology.[51] Gene drive technology makes it possible to release modified genomes via insects onto crops and even humans, with the aim of genetically modifying them in turn via cross-pollination or bites, respectively. The main funding institution of this highly controversial research is DARPA, but it was also discovered in a publicly available collection of documents that the Gates Foundation had paid a public relations firm to settle the discussion in the UN, where the criticism had been raised, in favor of the proponents.[52]

Other advanced microbiology research in the U.S. and China involved the use of nanomaterials (one nanometer is one billionth of a meter). A project at MIT on nanocrystals, funded by the Gates Foundation, also had a visiting researcher from the Institute of Chemistry in Beijing, Jing Lihong, working on it. These nanocrystals, so-called quantum dots, are tattooed under the skin and can over several years transmit a signal readable by a smart phone for the purpose of biometric identification and for access to the vaccination history and other data.[53] I will come back to this in the next chapter.

In 2013, Professor Charles Lieber on behalf of Harvard University had concluded an agreement with Wuhan University of Technology, officially for the nano-wiring of lithium-ion batteries. In reality, Lieber's laboratory was researching the integration of nanowires into living organisms. The batteries were merely a cover for this highly sensitive research. Nanowires are injected into the brain or retina (lens of the eye), where they wrap around neurons and record electrical communication between cells. In this way, it will ultimately be possible to penetrate into the sensory perception of the vaccinated animal or person. However, in January 2020, Lieber was arrested by the FBI.[54]

According to Francis Boyle, the aforementioned author of the U.S. bio-weapons legislation, Lieber was also involved in smuggling biological weapon materials, including the coronavirus, from the National Microbiology Laboratory (NML) at the University of Manitoba in Winnipeg, to China.[55] At the NML, the 'Fort Detrick' of Canada, Dr Xiangguo Qiu, a Chinese Canadian virologist, worked on an Ebola drug. Her husband, a biologist of Chinese nationality, worked with her. According to the controversial Israeli bioweapons expert Dany Shoham (he was among those who claimed that the anthrax attacks on U.S. politicians in 2001 were organized by Iraq), the project was supported by DTRA of the U.S. and Dr. Qiu also worked with three researchers from Fort Detrick, Maryland.[56]

In July 2019 it came to light that the NML in March, three months earlier, under the authority of Dr Qiu, had dispatched Ebola and Henipah viruses to Beijing. The U.S. CDC had classified these viruses as bio-terrorism agents because of their easy distribution, and the shipment did not have the required documentation to indicate that they were Canadian intellectual property. Dr Qiu, her husband, and a number of research students were subsequently denied access to the NML, something that had long been requested by Canadian intelligence services.[57] According to Shoham, all this was part of espionage for the Chinese bio-weapons program;

he claimed Wuhan is a bio-weapons laboratory, something actually confirmed by Francis Boyle.[58]

Of course, the Wuhan laboratory was also in the news following the SARS-2 outbreak. The oft-heard objection that laboratories have strict safety rules does not hold: after the first SARS outbreak, as a result of security protocols incorrectly drafted or inadequately enforced, that virus resurfaced in laboratories in Singapore, Taipei and Beijing.[59] Meanwhile the lab escape theory has become mainstream (see below). It had always been unlikely that a bat virus would have jumped to humans in the Wuhan market, because no bats were among the live animals for sale there.[60] In addition, the genome of a bat is so distant from that of humans that a virus cannot easily be transmitted between humans afterwards; that requires upgrading in a laboratory.

AN ARTIFICIAL VIRUS

In February 2020, Dr. Leonard Horowitz, the whistleblower fired by Henry Schein Corp. (a dental supply company also trading in vaccines), alerted U.S. authorities and the media that an Indian group led by Dr. Prashant Pradhan had determined the month before that the SARS-2 spike protein infection mechanism (by which it attaches itself to the ACE-2 receptor) contained four genes from the HIV1 virus, evidence that it had been developed in a laboratory. This was also Fourtillan's theory. However, the group, which in addition to Pradhan himself (head of the South Asia division at the IBM Watson artificial intelligence laboratory) consisted of nine renowned geneticists, was pressured to take back their report. Horowitz received no response from the authorities to whom he had addressed his warning and the media was also unwilling to take this up.[61] Apparently this was a possibility not even to be mentioned.

In fact, the inner circle of virologists involved in gain-of-function research had already acknowledged the lab theory. From the 3,000 Fauci e-mails made public following a Freedom of

Information Act (FOIA) request, it transpires that in the evening of 31 January 2020, one of these researchers, Kristian G. Anderson, sent an urgent email to Fauci and Jeremy Farrar, the director of the Wellcome Trust. Anderson thanked Fauci for sharing an article reporting the gain-of-function research the NIH/NIAID director had subcontracted to Wuhan via EcoHealth Alliance. The article, 'A SARS-like cluster of circulating bat coronaviruses shows potential for human emergence,' had appeared in *Nature Medicine* on 9 November 2015 as one of the aforementioned series co-authored by Dr Shi Zheng-li. Anderson then warned Fauci that the new SARS-2 virus looked suspicious, as 'some of the features look engineered' and this set in motion a mega-deception operation.[62]

Half an hour after midnight Fauci mailed his deputy to prepare for a busy day. Later that Saturday, 1 February, Farrar invited the gain-of-function set to take part in a confidential teleconference. Apart from Fauci and another NIH figure, Farrar and two other Wellcome officials, the participants were Marion Koopmans and Ron Fouchier of Erasmus MC Rotterdam, Christian Drosten of PCR test fame, and Patrick Vallance (head of research at GlaxoSmithKline and soon to become chief scientific adviser of the UK government), plus of course, Anderson. What was said in the conference was recorded in a flurry of follow-up e-mails, all heavily redacted in the Fauci/FOIA collection, although Koopmans and Fouchier do not meet the criteria that makes such redaction legal. However, on 9 February, Anderson and four co-authors (some of whom had also participated in the teleconference) published 'The Proximal Origins of SARS-CoV-2' in *Nature Medicine*. This was obviously intended to once and for all refute Anderson's own original warning that the virus was engineered, something he had dismissed as a 'crackpot theory' already in an email five days before, and the media duly had followed suit.[63]

Meanwhile the lab theory had begun to circulate. Still in February 2020, Steven Mosher, director of the Population Research Institute in Front Royal, Virginia, stated that the new virus had accidentally escaped from the laboratory in Wuhan. The Chinese ambassador to Washington, Ciu Tankai, called this an insane theory.[64] In China itself, researchers initially went no further than establishing that the virus the first patient was hospitalized with in Wuhan in December 2019, is closely related to SARS-like corona-viruses found in bats in China, and that its nucleotides were 89.1 percent identical with that group; however, they did not address what the non-identical 11.9 percent consisted of.[65] Other Chinese researchers found that the SARS-2 virus uses the same method of evading the human immune system as the actual HIV. Both viruses do this by removing a 'marker' molecule on the surface of an infected cell (the major histocompatibility complex, MHC), thus entering it unopposed and making the infection chronic. The spike protein with which this happens, ORF8, is not found in the SARS virus in bats.[66]

Now the voices calling this an artificial virus got louder, although their contention continued to be furiously dismissed by the media as 'conspiracy theory' for another one-and-a-half year. Ruan Jishou of Nankai University and Li Huan of Huazhong University independently concluded that SARS-2 was a laboratory product.[67] A Czech microbiologist, Dr Soňa Peková, caused a furore by refuting American claims that it was a natural mutation. She pointed out that the 'control room' at the beginning of the RNA helix, comparable to a computer's BIOS, had been reorganized in such a way that it was hard to imagine that this virus could survive at all.[68] Dr Judy Mikovits, a leading albeit too outspoken scholar (blackballed after her findings threatened the interests of the pharmaceutical industry and associated government agencies) stated that the number of mutations in SARS-2 was so high that nature would need 800 years or so to achieve anything like it.[69]

Early on, a decisive vote had already been cast by the French virologist, Luc Montagnier who, together with Françoise Barré-Sinoussi, had won the Nobel Prize for Medicine in 2008 for the discovery of the HIV virus (in 1983). Having determined with mathematician Jean Claude Perez that the SARS-2 virus contained sequences of the HIV genome in its non-identical part—HIV1, HIV2 and also malaria—Montagnier expressed his disappointment that research of Pradhan et al. had been withdrawn 'under political pressure.'[70] Montagnier made the comparison to a puzzle, in which the coronavirus has 30,000 pieces and HIV-1, HIV-2 and SIV (a retrovirus close to it) are puzzles of 9,000 pieces each. If three pieces of each of these last puzzles are neatly found next to each other in the big one, one may assume that this did not happen in nature.[71] Like Fourtillan, Montagnier believed that this mutation had been fabricated in the laboratory in Wuhan (he supposed in order to make an AIDS vaccine). Nature itself would remove this artificial mutation in due course, but before that the virus would make many more victims. As late as December 2020, vilified from all sides, Montagnier stuck to this statement.[72] Meanwhile others outside the microbiology community have also supported this view (Richard Dearlove, former MI6 chief on *Sky News* in April, stated that it was an 'engineered virus' that had escaped from Wuhan's lab). Of course, in the case of this veteran producer of fake news to justify the Iraq invasion, we have to be on guard for attempts to blame China and forget about U.S. direction of this research.[73]

Meanwhile those who knew about the true background of the engineered virus and their own responsibility (Fauci, Daszak), continued to trumpet the Anderson-2 version. Four days before the pandemic was declared by the WHO, an open letter including Daszak as one of the signatories was published in *The Lancet* (7 March 2020) warning against 'conspiracy theories' about the new virus. Daszak himself stated in mid-April that the new virus was a mix of two bat viruses and implied that it had somehow traveled from Yunnan to Wuhan (incidentally, the NIH discontinued the project with Wuhan a week later, on 24 April). Despite his earlier role and statements, Daszak was then given the task of investigating the origin of the virus with an international research team established by *The Lancet*. In January 2021, he traveled with a WHO delegation (including his co-conspirator, Marion Koopmans) to Wuhan to look there for the source of the infection.[74] Upon their return, the WHO stated that the delegation had not been able to establish that the virus originated in Wuhan and that the laboratory option was considered extremely unlikely. However, in June 2021

Daszak left the *Lancet* commission after his links to the Wuhan laboratory had been raised publicly.[75]

FROM FORT DETRICK TO WUHAN

The outsourcing of gain-of-function research on bat viruses to Wuhan is not the only possibility for the emergence of SARS-2. This research was also carried out in the U.S. itself, in Fort Detrick among other locations. As mentioned, it is routine when upgrading viruses in the laboratory to develop a vaccine against it at the same time, so that presumably, one is always protected. In 2018, CEPI, the vaccination network of Norway, India, the WEF, and the Gates Foundation, awarded a $56 million grant to the Inovio program to develop a vaccine against MERS; USAMRIID was a partner in this program. In the same year, a patent was granted (filed in 2015) for a vaccine containing an artificially manufactured coronavirus. According to the patent, it concerns an attenuated coronavirus with a gene that encodes proteins, and which can be used to fight or prevent diseases, such as infectious bronchitis.[76]

On 2 July 2019 an unknown respiratory disease was reported by a retirement home in Springfield, Virginia, some 50 miles from Fort Detrick. In retrospect, both the symptoms and the percentages of infected, sick and death were found to roughly correspond to the SARS-2 pattern (apart from the age structure, see below). The CDC could not identify the condition in the Springfield case, and the SARS-2 virus would not be identified until January 2020.[77] Instead, the respiratory disease, which soon showed up in 16 more sites, was attributed to vaping (smoking electronic cigarettes). However, in other countries where e-cigarettes are smoked, there was no question of it causing a respiratory disease.

By contrast the 'mysterious and life-threatening' vaping disorder in the U.S. spread quickly and became an epidemic. A physician cited in *The New York Times* warned that 'something is very wrong.'[78] Most patients suffered from breathing problems, chest pain, vomiting and fatigue, all symptoms of Covid-19, but they were mostly adolescents or young adults. Because the 2019–20 flu season in the U.S. also set in early and was unusually severe (according to a CDC estimate, more than 26 million Americans sick, 250,000 hospitalized, and at least 14,000 people died), it was difficult to separate the two conditions. Even so, several media already in February 2020 were adamant that the deadly respiratory virus was not the new coronavirus.[79] In Italy, one of the first countries badly affected by Covid-19, GPs noticed strange forms of pneumonia among the elderly as early as November 2019.[80]

Once the accusing finger was pointed at Wuhan, Chinese media did not remain silent. In March 2020 a Chinese newspaper pointed out the high number of flu patients in the U.S. that winter and reported that a petition filed on the White House website demanded clarity from the U.S. government on whether the closure of the laboratory at Fort Detrick had anything to do with the virus outbreak. In fact, as the Covid-19 pandemic began to take hold, many English-language news reports about the closure of Fort Detrick were removed from the Internet.[81]

The USAMRIID complex had indeed been closed by order of the CDC in July 2019 due to serious bio-safety deficiencies.[82] These in particular concerned the inadequate treatment of wastewater, because after a flood in May 2018, the steam sterilization installation that had been used previously was replaced by chemical decontamination. This was not the first time that Fort Detrick had been closed. In 2009, the year of the H1N1 outbreak (the Mexican or swine flu), this happened as well. At that time, it was about discrepancies in the stocks of infectious material. The reason for the 2019 closure was much more important because there was a risk of contamination, even though this was denied by spokespersons for Fort Detrick.[83]

The case hit the headlines after the U.S. team at the World Military Games in Wuhan (18–27 October 2019) put in an abysmal performance, partly due to illness. Of the 300 U.S. participants, some had trained shortly before departure at Fort Belvoir, Maryland, 50 miles from Fort Detrick but only 6 miles from Springfield, the site of the first outbreak of the unknown vaping respiratory disease. Once in China, the team stayed at the Wuhan Oriental Hotel, 300 meters from the Huanan fish market in Wuhan.[84]

When the U.S. National Security Adviser, Robert O'Brien, accused China of wasting valuable time by its slow response to the outbreak, the Chinese side did not hesitate to raise the issue of the closure of Fort Detrick.[85] The spokesman for the Chinese Foreign Ministry, Zhao Lijian, distributed an article from *Global Research*, the website of the Canadian Center for Research on Globalization, which claimed that the U.S. team had brought the virus to China. 'When did Patient Zero start in the U.S.? How many people are infected? What are the names of the hospitals?' Zhao Lijian asked in a Twitter post. Indeed, as early as 27 February, a leading Chinese epidemiologist, Dr Zhong Nanshan, had pointed out that 'although Covid-19 was first discovered in China, it did not mean the virus originated in China.'[86]

The New York Times rushed to dismiss the charge that the military team had brought the virus from the U.S. as a 'baseless conspiracy theory,'[87] but Zhao was not contradicted by his superiors, nor was the Twitter post retracted. And while honesty is the last thing a state should be suspected of, it is uncharacteristic of China's Foreign Ministry to publicly indulge in wild speculation. Diplomacy is the most conservative branch of any state apparatus and contender states consider sovereignty (and by implication, non-interference in internal affairs) as a cornerstone of foreign policy. A tweet from a spokesperson that may prove to be unfounded will not be left online if it might damage relations with China's most important partner in the world, although the Chinese ambassador to Washington, when asked about it, stuck to his earlier conclusion that the claim that the virus came out of a lab, was madness.[88]

In the meantime, an online discussion on this issue had begun in China and while it has no official status, the actual assertions made therein are plausible and worth considering.[89] Aside from the Olympic training near Springfield, Maryland and the staying at the Wuhan Oriental Hotel, near the fish market (Wuhan's laboratory, on the other hand, is 30 kilometers south of that hotel, with a river in between),[90] it further emerged from the Chinese Internet discussion that the American team went on to finish in 35th place, difficult to imagine under normal circumstances. In addition, on 25 October, some U.S. participants were admitted to an infectious disease hospital in Wuhan with a fever. Meanwhile, 42 employees of the Wuhan Oriental Hotel had been diagnosed with the condition that would later be referred to as Covid-19, the first cluster in Wuhan. Seven market vendors had just previously been diagnosed with the condition, and the hotel employees had been in contact with them.

So it is also possible that the infection trail ran the other way around, from the market to the hotel, and that the Americans were only infected there—but then the question remains why the 300-strong team of the largest army in the world performed so poorly. In addition, still according to Godfree Roberts' summary of the Chinese social media discussion, Secretary of State Pompeo phoned Yang Jiechi, China's State Councilor for Foreign Affairs, in mid-March, requesting the non-disclosure of what was said in China about the course of the infection. This was the highest level of communication, because normally Pompeo would have called Foreign Secretary Wang Yi (Yang is Wang's boss). Yang is said to have replied, 'We await your solemn explanation, especially on Patient Zero.'[91] All along, the tweet from the Foreign Ministry spokesman was not removed.

The relationship between the U.S. and China had already deteriorated significantly due to the U.S. trade measures and the allegations against Beijing about the virus. This was partly due to anti-Chinese pressure groups in the U.S. and Trump's refusal to follow the recommendation to implement a hard lockdown.

THE RED DAWN GROUP AND THE LOCKDOWN PLANS

I n the spring of 2020, Trump faced the challenge of handling the Covid pandemic in such a way that it would not jeopardize his re-election, which under normal circumstances would most likely have caused few problems for him. The intelligence-IT-media power bloc, as we saw, had already openly turned against him, mobilizing large swaths of the biopolitical complex, both military and civil, for an attack on the outsider president.

The proponents of a general lockdown per the European model, that is, creating panic followed by harsh measures, were organized in a group called *Red Dawn*—a striking choice of a name, recalling a 1984 feature film about a Soviet-Cuban invasion of the U.S., in which Nicaragua (!) also took part. In the U.S., the Red Dawn group wanted to carry out the scenario earlier conceived by Donald Rumsfeld, the architect, with Dick Cheney, of the modernization of the Continuity of Government blueprint for a state of emergency. In 1997 Rumsfeld had become Chairman of the Board of Directors of the pharmaceutical group Gilead Sciences, making him an insider in the biopolitical complex. After becoming Secretary of Defense to Bush Jr. in 2001, he had ordered Dr Richard Hatchett of the National Security Council and Dr Carter Mecher of the Department of Veterans Affairs to adapt the quarantine model developed for a biological attack on U.S. bases overseas, for use on the U.S. domestic civilian population. In the War on Terror, according to Rumsfeld, the military-civil distinction was no longer applicable anyway. In 2006, shortly before his retirement from the Pentagon, Hatchett and Mecher convinced the Centers for Disease Control to put this scenario into effect in case of a bio-medical emergency.[92]

When it turned out that Trump was reluctant to pursue the radical lockdown strategy, Mecher, Hatchett (meanwhile made head of the CEPI vaccination network), Fauci of the NIAID, Dr Robert Redfield of the CDC and about thirty more members of the Red Dawn network struck a major alarm. In emails to the group, Mecher demanded the closure of all educational institutions, because 'what was coming was unbelievable.' Hatchett was the first to characterize the Covid crisis as a war.[93] Cities the size of Chicago had to be placed under lockdown, to name but one of his proposals. Another member of the Red Dawn group predicted 490 thousand deaths in the U.S.[94]

Trump's health minister, Alex M. Azar II, joined the alarmists, as did his disaster department head, Dr Robert Kadlec. As we saw, Kadlec had been involved in previous exercises on the theme of bioterrorism, as early as the Dark Winter scenario of 2001; in 2019 he himself led the Crimson Contagion exercise, in which a virus originating from China was predicted to cause nearly 600,000 deaths in the U.S.[95] Once again, it is apparent to what degree Trump was an outsider in his own state apparatus. Contrary to what had happened in Europe and elsewhere, he did not want to hand over the actual power of government to the experts with their alarmist warnings. Anyone still remembering Trump's press conferences, with a smirking Fauci next to him, saw that this president was in fact being set up for ridicule by his advisers. His professed use of hydroxychloroquine as a prophylactic and other violations of the corona scenario were portrayed in the mainstream media as examples of a president who did not know what he was talking about and Trump's sneering at reporters facing him did the rest.

Whether the Covid scenario was used to reduce Trump's chances of re-election or perhaps even rolled out with that in mind, cannot be determined. But there is no doubt the president and Treasury Secretary Steven Mnuchin wanted to avoid anything that could aggravate economic disruption in the election year. That is why, despite his anti-Chinese rhetoric, Trump resisted pressure from the anti-Chinese group in the administration, led by Deputy National Security Adviser Matthew Pottinger, who wanted to hold Beijing directly liable for the virus and take punitive measures. Here too Trump appeared unwilling, but neither was he able to follow an alternative course of his own. As a result, an entry ban was imposed on travelers from China early on and Pottinger and

Secretary of State Pompeo persuaded the president to speak about the 'Wuhan virus' and to convince the G7 to follow suit.[96]

GLOBAL GOVERNANCE AND ONGOING RIVALRY

Tensions with China fit the more than a century old practice of contender states entrenching against the West, which has been more powerful throughout and routinely seeks to impose its will on any adversary. As in the early twentieth century, this antagonism risks an open conflict again today and one of the side effects of increasing rivalry is that the population becomes infected by the virus of chauvinism spread by the media. On the eve of World War I, the mainstream socialist workers' movement too was swept up in the mutual hatred encouraged by imperialist interests and their newspapers.

Karl Kautsky, the leader of German Social Democracy, commanded an audience much larger than left wing agitators like Rosa Luxemburg, who called for a general mobilization against the threat of war. Instead Kautsky suggested that the imperialists might strike a deal among themselves to jointly exploit the agricultural areas of the world (what would later be called the 'Third World') and avoid war altogether. Why not move beyond imperialism, and realize what he called an *ultra*-imperialism? Unfortunately for him, the article in which he asked this question came out just as World War I broke out.[97] But OK, perhaps after the war then?

Kautsky's rhetorical question (he already thought he saw a clear trend towards reconciliation) provoked a fierce tirade by the Russian revolutionary, Lenin, who saw Kautsky as the chief culprit in the willingness of the Social Democratic party leaders to sacrifice their supporters to the belligerency of the ruling classes in the European countries. Of course, Lenin argued, there is a trend towards a global trust in which all corporations and all states will be swallowed up. But that development is taking place under such great pressure, with such speed, and plagued by so many contradictions, conflicts, and shocks, that before it ever comes to that point, imperialism will be shattered.[98]

In the Covid crisis, this debate is again topical, as the question arises whether the exploiters of the world have finally reached the point where a synchronization of liberal capitalism and Chinese state capitalism is possible—or do Lenin's objections still hold? Are we witnessing the consolidation of a universal 'oligarchic collectivism,' as Orwell used the term—a truce between ruling classes holding their own populations captive in a nominal state of war?

After all, the situation today in many respects can be compared to the run-up to the First World War. 'It is undeniable that the coronavirus epidemic has come on the scene at a crucial moment, when people everywhere are in revolt against the power of international financial institutions and multinational pharmaceutical corporations, whose stranglehold on governments is no longer hidden,' writes the Belgian journalist, Senta Depuydt. 'Many scandals have shaken confidence. The bankruptcy of an aberrant economic system is accelerating and attempts to start a third world war are multiplying. While it is impossible to know how the "coronavirus pandemic" will influence the redistribution of power, it is certain that many are seeking to have Covid-19 serve the political interests of a global governance project.'[99] Of course, 'global governance' is little else but the contemporary equivalent of Kautsky's ultra-imperialism.

However, in the current situation, would China (and Asia generally) really accept to settle for participating with the West in an 'ultra-imperialism' when the West is evidently declining and plagued by disintegration? On the one hand, the answer is yes, namely on the issue of establishing an authoritarian state in which the formerly liberal homelands of capitalism adapt to the Chinese model by means of simultaneous domestic coups d'état leaving inequality untouched and political power in the hands of existing elites. All rulers in the world have an interest in assuring the acquiescence of their peoples, and in the foregoing chapters I have already indicated at length that this is the primary aim of the Covid state of emergency: that the current discipline is to be carried out under the pretext of a pandemic by means of a permanent state of emergency, to which popular majorities submit in the name of public health. For 40 years China has had to accept lessons from the West (economic, technological, on 'human rights'); now it is the West's turn for a lesson from the other side.

Yet would friction, conflict and even war really be eliminated as a result? As long as it only concerns the 'pandemic,' that is still conceivable. Everything here points to a cooperative 'ultra-imperialism' in Kautsky's sense, at least as far as the West is concerned. At a virtual meeting in early April 2020, for which invitations had been issued by the WHO, the President of France, the President of the European Commission, and the Bill & Melinda Gates Foundation, it was agreed to set up an Access to COVID-19 Tools Accelerator, or ACT Accelerator, to speed up the development, production and distribution of 'vaccines,' tests and therapy. 'Only solidarity will get us through this crisis,' WHO Director-General Tedros declared, citing the participating countries, health partners, manufacturers and the private sector, all of which were to benefit.[100]

The peak of unanimity seemed to have been reached in November 2020, when the G20 met, not in host country Saudi Arabia, but via Zoom. At the previous G20 in March, the world leaders attending had already translated the Covid emergency declared by the WHO into unanimous decisions instituting lockdowns and related measures. More than half a year later, this consensus was reaffirmed, with only Russian President Putin striking a note of dissent as he warned that unemployment, poverty and an unprecedented economic depression might be the greater danger.[101]

Still at the G20 meeting, the Chinese President, Xi Jinping, proposed to introduce global QR codes with biometric data, without which it should not be possible to travel. No one spoke out in favor of this proposal, nor against it either. The problem is that the West, Russia and China (and also Cuba) each have their own 'vaccines' and it depends entirely on whether they are mutually recognized before such a condition for travel can be universally imposed. As we will see, this has been no small source of potential conflict either, as intravenous gene therapies (only China's Sinovac vaccine deserves that name) are a source of unprecedented profit and the inoculation of healthy populations is so lucrative that its proceeds will not be left to rivals without a fight. There were plenty of other issues on which there was agreement, such as debt relief enabling poor countries to afford vaccines, which is another subsidy to the vaccine makers and also typical of Kautsky's reasoning: jointly exploiting the Third World. The EU, finally, proposed to focus on the World Economic Forum's program, the Great Reset, with its slogan 'Build Back Better.' Humanity has taken a big step towards unity at this summit, Israel Shamir conceded, adding however, that whilst this is better than disagreement, conflict, and war, the unity of leaders and experts is not yet unity of the peoples represented by them.

Indeed, the ongoing popular revolt is an inevitable source of disagreement, for a common policy in this area is practically impossible, given so many differences in circumstances. The temptation to make concessions at each other's expense is too great. In addition, there is the purely economic rivalry, since practically all countries are controlled by a capitalist oligarchy chasing maximum profits directly or indirectly (in Russia, China and Iran power is shared with a state class), but they are not equally firmly in the saddle everywhere.

Not only is there rivalry between the U.S. and China, but also between the U.S. and the EU regarding China. In January 2021, for example, after consultations between Xi Jinping, Merkel and Macron, the EU concluded an investment treaty that nominally also increases the opportunities for European companies in China, but that primarily gives Chinese companies access to the EU.[102] That this treaty, which had been negotiated for seven years, was finalized just before Biden's assumption of power, has everything to do with the plans of the new administration in Washington to switch to a harder line against Beijing.[103] With an extremely unstable international economic situation, with a depression approaching as a result of the lockdowns, that is almost a recipe for war—the only area in which the U.S., with NATO, still believe they have a chance of crushing China.

This brings us to the profit opportunities the 'pandemic' creates for the pharmaceutical industry and how these are intertwined with the IT power bloc's plans to place the world's population under permanent surveillance (*'leave no one behind'*).

Endnotes

1. Wang Hui, *The End of the Revolution: China and the Limits of Modernity* [trans. R. Karl *et al.*] (London: Verso, 2009), pp. 27–32.

2. F. William Engdahl, 'The Eurasian Century is Now Unstoppable: The transfer of the geopolitical center of gravity to Eurasia is something the West will have to get used to,' *Russia Insider*, 4 October 2016 (online).

3. See my 'Who Will Protect the Baltic Energy Highway?' in Van der Pijl, ed. *The Militarization of the European Union* (Newcastle: Cambridge Scholars, 2021), pp. 104–7.

4. David Lane, 'Post-Soviet Regions. From Interdependence to Countervailing Powers?' in David Lane and Guichang Zhu, eds., *Changing Regional Alliances for China and the West* (Lanham, Maryland: Lexington Books), 2018, pp. 6–7.

5. Zbigniew Brzezinski, *Strategic Vision: America and the Crisis of Global Power* (New York: Basic Books, 2013 [2012]), pp. 4, 75–89.

6. John Lee, 'China Steps up the Long March to 5G: As the U.S. increases pressure on technology exports, China's race for self-sufficiency becomes more urgent,' *The Diplomat*, 6 May 2020 (online).

7. Eamonn Fingleton, *In Praise of Hard Industries: Why manufacturing, not the information economy, is the key to future prosperity* (Boston: Houghton Mifflin, 1999), pp. 117–23; Lee, 'China Steps up the Long March to 5G.'

8. Gisela Grieger, '5G in the EU and Chinese telecoms suppliers,' *European Parliament Research Service, At a Glance*, April 2019 (online); *Deutsche Welle*, 'EU rules out Huawei ban but maps out strict ruiles on 5G,' 28 January 2020 (online).

9. *Great Game India*, 'Transcript: Bioweapons Expert Dr. Francis Boyle on Coronavirus,' 5 February 2020 (online).

10. Max Parry, 'Is the Global Pandemic a Product of the Elite's Malthusian Agenda and U.S. Biowarfare?,' *Unz Review*, 16 March 2020 (online).

11. Pierre Lescaudron, 'Compelling Evidence that SARS-CoV -2 Was Man-Made,' *Sott.net*, 26 June 2020 (online).

12. Whitney Webb, 'Bats, Gene Editing and Bioweapons: Recent Darpa Experiments Raise Concerns Amid Coronavirus Outbreak,' *Unz Review*, 30 January 2020 (online).

13. Stefan Elbe, *Security and Global Health. Towards the Medicalization of Insecurity* (Cambridge: Polity Press, 2010), p. 73.

14. Patrick Zylberman, *Tempêtes microbiennes. Essai sur la politique de sécurité sanitaire dans le monde transatlantique* (Paris: Gallimard, 2013), pp. 95–97.

15. Webb, 'Bats, Gene Editing and Bioweapons.'

16. Lescaudron, 'Compelling Evidence That SARS-CoV-2 Was Man-Made.'

17. *Wikipedia*, 'Sidney Gottlieb'; Wayne Madsen, 'U.S. conducted biological weapons research at least until 2003 despite 1969 presidential order, 1972 treaty,' *Strategic Culture*, 23 August 2016 (online).

18. Parry, 'Is the Global Pandemic a Product of the Elite's Malthusian Agenda?'

19. Lescaudron, 'Compelling Evidence That SARS-CoV-2 Was Man-Made'; *Al Jazeera World*, 'Who Killed Robert Kennedy?' 14 November 2018 (online).

20. *Wikipedia*, 'MKNAOMI'; Charles Morgan, 'What's in the Future for the National Security Community?' *West Point Visual Information Division*, 17 April 2018 (online).

21. Elbe, *Security and Global Health,* pp. 73–74.

22. Madsen, 'U.S. conducted biological weapons research.'

23. Judy Mikovits and Kent Heckenlively, *Plague of Corruption. Restoring Faith in the Promise of Science* [foreword Robert F. Kennedy, Jr.] (New York: Skyhorse, for Children's Health Defense, 2020), p. 61.

24. *Great Game India*, 'Transcript: Bioweapons Expert Dr. Francis Boyle'; Zylberman, *Tempêtes microbiennes*, p. 143.

25. Webster G. Tarpley, *9/11 Synthetic Terror Made in USA*, 4th ed. (Joshua Tree, Calif.: Progressive Press, 2008 [2005]), pp. 311 ff.; Elbe, *Security and Global Health,* pp. 77–8.

26. Parry, 'Is the Global Pandemic a Product of the Elite's Malthusian Agenda?'; Madsen, 'U.S. conducted biological weapons research.'

27. Zylberman, *Tempêtes microbiennes.* pp. 281–82, 286.

28. Elbe, *Security and Global Health,* p. 81; Webb, 'Bats, Gene Editing and Bioweapons.'

29. Madsen, 'U.S. conducted biological weapons research.'

30. Dilyana Gaytandzhieva, 'The Pentagon Bio-weapons,' *Dilyana.Bg*, 29 April 2018 (online).

31. Webb, 'Bats, Gene Editing and Bioweapons.'

32. Elbe, *Security and Global Health*, p. 75.

33. Gaytandzhieva, 'The Pentagon Bio-weapons.'

34. Madsen, 'U.S. conducted biological weapons research.'

35. Gaytandzhieva, 'The Pentagon Bio-weapons.'

36. Ibid.

37. Judy Mikovits considers Ebola (like HIV) the result of vaccination using animal tissue; Mikovits and Heckenlively, *Plague of Corruption*, pp. 152–58.

38. Webb, 'Bats, Gene Editing and Bioweapons.'

39. Ibid.; MERS fatality figures in Walter Van Rossum, *Meine Pandemie mit Professor Drosten. Vom Tod der Aufklärung unter Laborbedingungen* (Neuenkirchen: Rubikon, 2021), p. 85.

40. Michael S. Northcott, email 30 January 2021, in Northcott and Daniel Broudy email exchange on Propaganda and the 'War on Terror' (online).

41. See the Erasmus MC website, e.g., 'Principal investigator: Prof. R.A.M. (Ron) Fouchier, Deputy Head of the Erasmus MC Department of Viroscience' (online).

42. Gaytandzhieva, 'The Pentagon Bio-weapons.'

43. Lescaudron, 'Compelling Evidence that SARS-CoV-2 Was Man-Made.'

44. Steve Hilton, 'Steve Hilton investigates origins of COVID-19, links to U.S. commissioned research,' *The Next Revolution (Fox News)*, 25 January 2021 (online); Madsen, 'U.S. conducted biological weapons research.'

45. Cited in Katherine Eban, 'The Lab-Leak Theory: Inside the Fight to Uncover COVID-19's Origins,' *Vanity Fair*, 3 June 2021 (online).

46. Anthony Fauci, Gary Nabel and Francis Collins, 'A Flu Virus Risk Worth Taking,' *The Washington Post*, 30 December 2011 (online); Hilton, 'Steve Hilton investigates origins of COVID-19,' *Washington Post*, 30 December 2011.

47. Flo Osrainik, *Das Corona-Dossier. Unter falscher Flagge gegen Freiheit, Menschenrechte und Demokratie* [foreword Ullrich Mies] (Neuenkirchen: Rubikon, 2021), p. 195.

48. Fred Guterl, 'Dr. Fauci Backed Controversial Wuhan Lab with Millions of U.S. Dollars for Risky Coronavirus Research,' *Newsweek*, 28 April 2020 (online); Hilton, 'Steve Hilton investigates origins of COVID-19.'

49. One Richard A. Rothschild applied for a patent on a Covid test in 2015 and again one in 2020. 'Accomplished pharma prof thrown in psych hospital after questioning official Covid narrative,' *LifeSite*, 11 December 2020 (online); 'Le Pr Fourtillan, un des intervenants du documentaire *Hold-Up*, place en détention provisoire,' *Le Figaro*, 18 April 2021 (online).

50. Guterl, 'Dr. Fauci Backed Controversial Wuhan Lab'; Hilton, 'Steve Hilton investigates origins of COVID-19.'

51. Webb, 'Bats, Gene Editing and Bioweapons'; Arthur Neslen, 'U.S. military agency invests $100m in genetic extinction technologies,' *The Guardian*, 4 December 2017 (online).

52. *ETC Group*, 'The Gene Drive Files: Disclosed emails reveal military as top funder; Gates Foundation paying $1.6 million to influence UN expert process,' 4 December 2017 (online).

53. The original link has been removed but I discovered it on *MIT Africa.*, 'Storing medical information below the skin's surface,' 3 February 2020, (online).

54. Robert F. Service, 'Why did a Chinese university hire Charles Lieber to do battery research?,' *Science,* 4 February 2020 (online).

55. *Great Game India*, 'Transcript: Biowapeans Expert Dr. Francis Boyle.'

56. Dany Shoham, 'China and Viruses: The Case of Dr. Xiangguo Qiu,' *BESA Center Perspectives Paper* No. 1,429 (The Begin-Sadat Centre for Strategic Studies), 29 January 2020 (online)

57. Nicoletta Lanese, 'Questions surround Canadian shipments of deadly viruses to China,' *The Scientist*, 9 August 2019 (online).

58. Shoham, 'China and Viruses'; *Great Game India*, 'Transcript: Bioweapons Expert Dr. Francis Boyle.'

59. Elbe, *Security and Global Health*, p. 41.

60. Fan Wu *et al.*, 'A new coronavirus associated with human respiratory disease in China,' *Nature*, vol. 579, 3 February 2020, pp. 265–269 (online).

61. Leonard G. Horowitz, *Complaint for Injunctive Relief Against Unfair and Deceptive Trade by Civil Conspiracy in Violation of the Florida Whistleblower Act, Civil Rights, and Public Protection Laws*. Civil suit against Pfizer, Inc., Moderna Inc., Hearst Corp. and Henry Schein, Inc., U.S. District Court for Middle Distric of Florida, 1 December 2020 [pdf], pp. 20, 28; Prashant Pradhan *et al.*, 'Uncanny similiarity of unique inserts in the 2019-nCooV spike protein to HIV-1 gp120 and Gag' [withdrawn], *BioRXiv: The Prepint Server for Biology*, n.d. [2020] (online).

62. I follow the reading of the Fauci e-mails and events by Chris Martenson, 'Fauci Lab Leak Cover-up Revealed Via Emails,' *Peak Prosperity*, Episode 007, 5 June 2021 (online).

63. Ibid.

64. Jonathan Swan and Bethany Allen-Ebrahimian, 'Top Chinese Official Disowns U.S. Military Lab Coronavirus Conspiracy,' *Axios on HBO*, 22 March 2020 (online).

65. Fan Wu *et al.*, 'A new coronavirus associated with human respiratory disease.'

66. Tyler Durden, 'Coronavirus Uses Same Strategy As HIV To Evade, Cripple Immune System: Chinese Study Finds,' *Zero Hedge*, 27 May 2020 [orig. *South China Morning Post* same date] (online).

67. Lescaudron, 'Compelling Evidence That SARS-CoV-2 Was Man-Made.'

68. *Dimsumdaily Hong Kong,* 'Czech molecular biologist, Dr. Soňa Peková explains in layman terms that COVID-19 virus originates from a lab, and the Americans try to refute it' (updated), 30 March 2020 (online).

69. Judy Mikovits cited in Lescaudron, 'Compelling Evidence That SARS-CoV-2 Was Man-Made.'

70. Horowitz, *Complaint for Injunctive Relief Against Unfair and Deceptive Trade*, p. 29.

71. 'Accomplished pharma prof thrown in psych hospital,' *LifeSite*.

72. Jean Claude Perez and Luc Montagnier. 'Covid-19, SARS and Bats Coronaviruses Genomes Peculiar Homologous RNA Sequences,' *International Journal of Research—GRANTHAALAYAH*, 8 (7), 2020, pp. 217–63 (online version); *FranceSoir*, 'Le défi de la vérité: Luc Montagnier, prix Nobel de médicine,' 17 December 2020 (online).

73. Emily Makowski, 'Theory that Coronavirus Escaped from a Lab Lacks Evidence. The pathogen appears to have come from wild animals, virologists say, and there are no signs of genetic manipulation in the SARS-CoV-2 genome,' *The Scientist*, 5 March 2020 (online); the Dutch 'newspaper of record,' *NRC-Handelsblad*, smeared the expert truth-tellers on 20 April 2020 under the heading, 'Corona Jihad.'

74. Hilton, 'Steve Hilton investigates origins of COVID-19.'

75. Anne Gulland, 'UK scientist with links to Wuhan lab "recuses himself" from inquiry into Covid origins,' *The Telegraph*, 22 June 2021 (online); Hilton, 'Steve Hilton investigates origins of COVID-19.'

76. Lescaudron, 'Compelling Evidence That SARS-CoV-2 Was Man-Made.'

77. Ibid.

78. Sheila Kaplan and Matt Richtel, 'The Mysterious Vaping Illness That's Becoming an Epidemic: A surge of severe lung ailments has baffled doctors and public health experts,' *The New York Times*, 31 August 2019 (online).

79. Dennis Thompson, 'Flu Season That's Sickened 26 Million May Be at Its Peak,' *U.S. News*, 21 February 2020 (online).

80. Godfree Roberts, 'Last Man Standing: China Wins Big With Covid-19. What Were We Thinking?,' *Unz Review*, 22 March 2020 (online).

81. Leng Shumei, 'Why was a U.S. military lab handling high-level disease shut down in July 2019?,' *Global Times*, 15 March 2020 (online).

82. Webb, 'Bats, Gene Editing and Bioweapons.'

83. Shawna Williams, 'CDC suddenly shuts down U.S. Army's Fort Detrick bioweapons lab due to "lapses in safety,"' *The Scientist*, 7 August 2019 (online); Parry, 'Is the Global Pandemic a Product of the Elite's Malthusian Agenda?'

84. Lescaudron, 'Compelling Evidence That SARS-CoV-2 Was Man-Made'; Sarah M. McClanahan, 'Innovation Fitness—Soldiers Push Their Limits.' *Maryland National Guard News*, 3 August 2019 (online).

85. *The Straits Times,* 'U.S. military may have brought coronavirus to Wuhan, says China in war of words with U.S.' (13 March 2020, online).

86. 'China Focus: Confident novel coronavirus outbreak under control by late April: Health expert,' *Xinhua.net*, 27 February 2020 (online).

87. Steven Lee Myers, 'China Spins Tale That the U.S. Army Started the Coronavirus Epidemic,' *New York Times*, 13 March 2020 (online).

88. Swan and Allen-Ebrahimian, Top Chinese Official Disowns U.S. Military Lab Coronavirus Conspiracy.'

89. Roberts, 'Last Man Standing: China Wins Big With Covid-19,' based on Thomas Polin Hon Wing, 'Wuhan's Outbreak: China Demands An Honest Accounting,' *Facebook,* 1 March 2020.

90. Frances Mulraney and Glenn Owen, 'Revealed: U.S. government gave $3.7 million grant to Wuhan lab at center of coronavirus leak scrutiny that was performing experiments on bats from the caves where the disease is believed to have originated,' *Mail on Sunday,* 12 April 2020 (online).

91. Cited in Roberts, 'Last Man Standing: China Wins Big With Covid-19.'

92. Thierry Meyssan, 'Covid-19 and The Red Dawn Emails,' *Voltaire Network*, 28 April 2020 (online); Pepe Escobar, 'How Biosecurity is Enabling Neo-Feudalism,' *Strategic Culture Foundation*, 15 May 2020 (online)

93. Meyssan, 'Covid-19 and The Red Dawn Emails'; Patrick Henningsen, 'COVID-19: Trigger for a New World Order: Economic Stagnation and Social Destruction,' *Global Research*, 24 August 2020 (online).

94. Matthew Mosk, Kathlyn Folmer en Josh Morgoilin, 'As Coronavirus threatened invasion, "Red Dawn" team tried to save America,' *ABC News*, 28 July 2020 (online).

95. Eric Lipton *et al.*, 'He Could Have Seen What Was Coming: Behind Trump's Failure on the Virus,' *The New York Times*, 11 April 2020 (online); Webb, 'All Roads Lead to Dark Winter.'

96. Lipton *et al.*, 'He Could Have Seen What Was Coming.'

97. Karl Kautsky, 'Der Imperialismus,' *Die Neue Zeit*, vol. 2 (1914), pp. 908–22.

98. See Lenin's preface to Nikolai Bukharin, *Imperialism and World Economy* (London: Merlin, 1972 [1915]), and his own subsequent *Imperialism, the Highest Stage of Capitalism* [1917] in *Collected Works*, vol. 22 (Moscow: Progress, 1964).

99. Senta Depuydt, 'Does the coronavirus pandemic serve a global agenda?,' *Sott.net* [orig. *Children's Health Defense.org*], 20 March 2020 (online).

100. *WHO*, 'Global leaders unite to ensure equitable access new vaccines, tests and treatments for Covid-19,' 4 April 2020 (online).

101. Israel Shamir, 'Eat Your Lemon!' *Unz Review*, 26 November 2020 (online).

102. Jack Denton, 'Biden's China priorities could be challenged by the landmark new EU-China investment treaty,' *Market Watch*, 1 January 2021 (online).

103. See Joseph R. Biden, Jr., *Renewing America's Advantages: Interim National Security Strategic Guidance* [pdf] (Washington: The White House, March 2021).

6.

THE 'PANDEMIC' AS
DISASTER CAPITALISM

We have seen that following the collapse of the Soviet bloc and the USSR, a moral and political crisis began to manifest itself in the West. Signs of popular unrest and even insurrection multiplied from the turn of the millennium and especially after the financial collapse of 2008. Of the scenarios for a response to these challenges, the option of a 'pandemic' with the corollary state of emergency was put on the agenda at an early stage. The biopolitical complex, mentioned a few times already and connected to the dominant U.S. intelligence-IT-media bloc via Bill Gates and the Bill & Melinda Gates Foundation, Michael Bloomberg and the health institutes at Johns Hopkins University, and others, in these circumstances adopted a strategy of predatory enrichment which has been labeled disaster capitalism.

The role of the pharmaceutical companies within this complex is best understood as a parasitic factor and in fact, with this sector we enter the outright criminal sphere, which is not a polemical exaggeration. The pharmaceutical companies are without exception repeat offenders when it comes to issuing false information about medications, including vaccines and related therapies. When the path of a pandemic came out on top in the elites' quest for a nightmare scenario that would enable population control, and the WHO with its newly acquired formal authority assumed a leading role in it, the pharmaceutical companies joined the fray, as they had previously done in the Mexican/swine flu epidemic.

I begin with the role of the Gates Foundation and then pursue how the coalition behind the pandemic scenario formed in the course of 2018–19, well before 'the virus' became public. We then turn to the experimental genetically engineered vaccines that are being developed which, albeit within a radically shortened trial trajectory, are to usher humanity into a new era. The original aim of permanent surveillance of a restive world population here acquires a biopolitical twist. The accompanying corruption leaves all previous pharmaceutical companies' scandals far behind. The illusion has even taken hold that undesirable political developments can be prevented biochemically, a wildly premature but yet noteworthy development that will direct the course of future research if left unchecked. The chapter concludes with a brief review of the visions of the future that have been developed about an artificially intelligent humanity—before returning, in the final chapter, to the roads to humanity's achievement of an active, authentic intelligence without which a humane future is inconceivable.

THE BIOPOLITICAL POWER COMPLEX AND THE GATES FOUNDATION

Within the force-field that led the world into a state of emergency in early 2020, the large pharmaceutical companies and biotech companies represent what Naomi Klein in her book, *The Shock Doctrine,* calls 'disaster capitalism,' documenting how, after a catastrophic crisis, under the heading of 'reconstruction,' radical state privatization policies open the doors for corporations to descend upon the wreckage, sweeping the detritus of the emergency into a source of profit.[1]

It is not difficult to see this mechanism at work in the Covid crisis as well. Disaster capitalism here consists in the opportunities that are created for the pharmaceutical industry and the financiers behind it. In Europe, hospitals and the related health industry are not privately owned to the same degree as in the U.S., but the state is nonetheless retreating further and further from its role in this domain. Actual privatization is partly replaced, partly supplemented by the abolition of the heretofore standard testing process for medication and vaccine development. This can be seen clearly in the case of the experimental gene therapies that are being pressed upon the public in great haste, without it having been determined, let alone publicly debated, whether this is warranted. The IT sector and the media back this criminal enterprise by suppressing objective information, including experts' critical scrutiny, and in general, negative public reaction that could potentially get in the way of enormous sales of untested pharmaceutical products.

We have already seen that in 2005 the WHO presented a document to the member states setting out their obligations in the event of a health crisis. One of these obligations was to purchase medicines and vaccines under preliminary 'sleeper' contracts as soon as the health organization would declare a pandemic. At the time, a pandemic denoted a cross-border epidemic of a new type causing large numbers of infections and deaths. But in 2009, that definition was revised downward. For Novartis, Hofmann-La Roche, Baxter Vaccines, GlaxoSmithKline, Sanofi Pasteur, and other giant pharmaceutical companies, these contracts have been a goldmine that has helped make the pharmaceutical industry one of the sectors with the highest profitability in today's economy.

The Covid state of emergency has all the features of biopolitics as defined by the French philosopher, Michel Foucault, a concept that I have used several times already.[2] In his lectures in the 1970s Foucault had pointed out that power, including the power to impose a state of emergency, was no longer primarily exercised at the level of the nation-state. It had moved to the international level, at the interface between a geographical, climatic and physical environment on the one hand, and the human species endowed with a body and a soul on the other. Here the sovereign will have to exercise its power at the intersection 'where nature in the sense of the physical elements interacts with nature in the sense of the condition the human species.' And then he concludes, in a chilling passage: 'Here the sovereign will want to intervene *if he wants to change the human species.*'[3] Foucault does not usually specify who would be exercising this 'will,' because he tends not to dwell on the ruling class role when presenting his insights about disciplining the population. However, in the current Covid crisis, there is no doubt that this is now a global project of which the individual states are mostly the relays. National governments and expert bodies cannot therefore be expected to be fully literate on all the details of the original authoritarian scenario. This knowledge does exist, but not as a single master plan. It circulates, in more or less complete versions, through the various transnational networks and there achieves a measure of cohesion and consensus. There is never a single source, no 'Schwab' who invented it all and directs the whole process from behind the scenes. The founder and animator of the World Economic Forum is at most the secretary of the project, or better, a series of overlapping projects.

The Bill & Melinda Gates Foundation, founded in 2000 and already mentioned several times, is not the all-powerful *deus ex machina* either, but its role in the biopolitical complex is paramount. The biopolitical complex further encompasses (apart from the pharmaceutical industry) the actual health and complementary care sector, the international and national organizations dedicated to public health such as the WHO, the World Bank health division, national public health institutions such as the CDC and the NIH in the United States and their equivalents in other countries, as well as NGOs such as Doctors Without Borders.

The central focus for Gates and his Foundation is the global market for synthetic medication, especially the most advanced therapeutic technologies. Like no other, Gates is a proponent of the biochemical approach to disease and the medicalization of social problems. In this regard, the Gates Foundation continues the tradition of the Rockefeller Foundation, which on the eve of World War I was instrumental in shaping modern medical science based on the allopathic approach. Here it should not be forgotten that as noted, medicines have climbed to the third cause of death in the U.S. and Europe, after cardiovascular disease and cancer,[4] to say nothing of their misuse, as evidenced by the vast opioid crisis in the United States.

Worldwide vaccination campaigns have been Gates's main preoccupation, wherein vaccine development is projected as by far the shortest route to public health, much more than by improving living conditions or strengthening preventative health care. It is certainly the most profitable. The Global Alliance for Vaccines and Immunization (GAVI) received a start-up grant of $750 million from Gates when it was founded in 1999, and from 2000 to 2014, the Gates Foundation contributed an additional quarter ($2.4 billion) of its total budget. In addition, Gates managed to get rich states such as Germany and others to contribute large amounts.[5]

At the World Economic Forum in Davos in 2010, Gates called for making the next 10 years the 'Vaccine Decade' and pledged $10 billion to that end. In December of the same year, a 'Global Vaccine Action Plan' was formulated for which Gates also enlisted the WHO, the UN Children's Fund (UNICEF), and Dr Fauci. Fauci was also on the project's Leadership Team alongside Anthony Lake of UNICEF, Margaret Chan, head of the WHO, and others.[6]

Due to their wealth and targeted subsidies, Gates and his foundation have an unparalleled influence, approximated only by the second richest, the Wellcome Trust. From 2004 onwards, the Gates Foundation's annual health allocations exceeded the WHO budget.[7] Martens and Seitz established that foundations such as Gates's are giving rich countries and their businesses a free hand in developing countries by establishing public-private partnerships with pharmaceutical companies and agribusiness. Biotechnology (genetic engineering) plays a major role in this, both in health care and in agriculture. In this light, the links of the Gates Foundation and the pharmaceutical industry are important as well. The many double appointments and subsidy and investment patterns of the biopolitical complex are reminiscent of the Military-Industrial Complex. Gates Foundation directors often come from major drug manufacturers such as Merck, GlaxoSmithKline, Novartis, Bayer HealthCare Services and Sanofi Pasteur. The Gates Foundation in turn provides grants to several of these companies, but it also has shares in them, so these grants are partly donations to its own assets. In addition, the Foundation gives money to pressure groups such as the Drug Information Association (run by the pharmaceutical industry) and the International Life Sciences Institute (funded by the major agribusiness companies). In 2006, Gates founded an alliance for a green (GM) revolution in Africa (AGRA), with the Rockefeller Foundation.[8]

The influence Gates has bought in the media was already discussed in Chapter 3. In 2015, the Gates Foundation also gave nearly $400,000 to the Poynter Institute to scour the Internet for unwanted health information and run purported fact-checks countering it. In its PR work the Poynter Institute lavishes praise on the McChrystal Group of the former JSOC commander. We saw already that the McChrystal Group is the leading cyber-propaganda company in the U.S., advising cities and states on lockdowns and handling PR for large pharmaceutical companies. In the 2020 election campaign the McChrystal Group was active in debunking Trump's statements about Covid through DefeatDisinfo.org. In addition, it campaigns against corona skeptics and 'anti-vaxxers.'[9]

Although Gates bragged in public lectures in 2013 and 2014 about the purported number of lives his work saved through partner agreements with the pharmaceutical industry (in one speech, he mentioned 10 million, in another, 6 million), according to Tim Schwab, his dual role as investor and donor has in fact made him an obstacle to attempts to make medicines accessible to the poor, in view of the intellectual property rights protection afforded to his patents on expensive medication.[10]

Gates also made statements comparing epidemics to war at a conference in Vancouver in 2015, where he spoke about the Ebola crisis in West Africa. There he complained that 'we' are not prepared for the next epidemic.[11] If there is one thing that will kill more than 10 million people in the coming decades, he warned, it will most likely be a highly contagious virus, rather than a war—not missiles, but microbes. In an interview, he also stated that he was working with the Wellcome Trust and MasterCard to develop antivirals, both vaccines and other therapies. What MasterCard has to do with the development of antiviral agents is best understood in light of its role as an interface between the bio-political complex and the IT-media bloc of forces that has seized power in the Covid crisis. By way of planning for pandemics, it became clear that Gates and the biopolitical complex wanted to seize the opportunity to set in motion a comprehensive, expensive production of test kits, antivirals and mRNA vaccines to sell to governments. Keeping a vulnerable global population engaged in a testing and treatment regime to be repeated over and over again by way of yearly boosters, etc. would then produce a vast captive market as it became inscribed into the 'new normal.'[12] Furthermore, this would largely sidestep the primary healthcare sector, and with it, the long-standing medications which have been found to be effective against Covid-19. We may therefore conclude that what we see in operation here is not just a business project but an all-encompassing philosophy in which humanity is conceived as a biomass of 7.5 billion specimens which, like agricultural crops, can be managed with genetic engineering, controlled and upgraded according to the principles of eugenics. Vaccination is therefore especially important for gaining access to the human organism, which in the scheme of things overall may render protection against actual disease as a concern of a lesser order.

While the WHO is so dependent on the conditional donations of the Microsoft founder—he is, after all, astronomically rich and received as if a head of state around the world, heading a founda-tion of widespread influence—it remains the only health organi-zation holding exclusive regulative authority under international law. Only the WHO can legally enforce the transmission of the preferences of Gates and the biopolitical complex in which he is the dominant force, on to the member states. It alone can prescribe binding policy on the basis of the treaty obligations that the states entered into in 2005 and ratified in subsequent years. That is why Gates cannot ignore the WHO, because the sovereignty of capital is only formally expressed via its post-2005 authority.

Until Trump withdrew the U.S. from the WHO, the Gates Foundation was the World Health Organization's second largest source of funding. For a brief period, Gates was the biggest (talk of regulatory capture!) until Biden reversed this decision. Fauci, the principal biopolitical officer within the U.S. state apparatus with an upgraded role under Biden, greeted this decision on camera by sending congratulations to 'his friend' Dr Tedros at the WHO. Yet Gates remains the main protagonist. As early as 2008, the head of the WHO's malaria program complained to Margaret Chan in an internal memorandum that Gates's influence was inhibiting scientific development by discarding divergent opinions and maintaining the homogeneous line put forward by the Gates Foundation, which 'could have dangerous consequences for the global public health policy process.'[13] In the years following the financial crash, the influence of Gates and his foundation has only increased, with more and more interests joining the nucleus he had established, poised and ready for 'disaster capitalism' opportunities on the health front.

CONVERGENCE OF INTERESTS BEHIND THE PANDEMIC SCENARIO

Bill Gates and Microsoft thus form the link between, on the one hand, the U.S. national security complex, the privatized IT industry plus the billionaires behind it, and the media; and on the other, the biopolitical complex around the Gates Foundation and the pharmaceutical industry. The ideas on the most appropriate responses to the challenges of what will likely be a highly restless, discontented world population, a possible new financial collapse and the rise of China, were gradually taking shape as a scenario such as Zelikow had formulated at the Miller Institute in 1998: the emergence of a public myth that comes about via a shock, a formative event, one that is shared and propagated by the relevant political community (the intellectuals speaking for the oligarchy and the urban cadre), and which is believed to be true by the masses.

In one respect, such a scenario develops very much like an 'organic' concept of control in that its components are already circulating in think tanks and planning bodies like the Trilateral Commission, the World Economic Forum, and the like. There, a synthesis is developed between the various interests at stake and a uniform narrative, a storyline, is created that makes the desired course seem logical and self-evident. Clearly the establishment of a public myth through a shock-like event requires a faster, express route through these networks, but otherwise the transnational coalition behind the pandemic scenario formed in much the same way as before. The WHO, the World Bank, the European Union, and subsidiary or otherwise related international organizations are thus again bound together as a quasi-state at the international level. Their connection to the United States as the leading state in the West and in the world is different in each combination, in the sense that various forms of sovereignty, private and public, are combined into one complex, layered structure, centered on transnational capital and the United States.

In 2016, at the World Economic Forum in Davos, the Innovation for Uptake, Scale and Equity in Immunization (INFUSE) project, developed by GAVI, the vaccination pressure group of the Gates Foundation, was set up. How it was connected to the harvesting of biodata and vax passes we will see later.[14] The following year, the aforementioned Coalition for Epidemic Preparedness Innovations (CEPI) was established by the governments of Norway and India, the World Economic Forum and again the Gates Foundation.

In 2018, on the occasion of the centenary of the Spanish Flu, the World Economic Forum, along with a number of other relevant networks, organized a meeting dedicated to combating the pandemic that was expected *with certainty*. The world was said to be ill-prepared for this, destined to arrive not so much as a result of health care cuts, war, or air and water pollution, but because of 'increasing trade, travel, population density, human displacement, migration and deforestation ... plus climate change.' It was, in short, an inevitability, and as a result, 'a new era of epidemic risk had begun.'[15] Happily, this new era had also dawned for the pharmaceutical industry, which, thanks to 'revolutionary new biotechnologies,' was now seen as capable of coping with emerging pandemics—an urgent need, since a major influenza pandemic alone could cost the global economy half a trillion dollars.

The WEF meeting on the Spanish flu was of course attended by the Gates Foundation itself, the institutions that the Gates Foundation had put on their feet (GAVI, CEPI, and the International AIDS Vaccine Initiative set up jointly with the Rockefeller Foundation in 1994) and of course, the pharmaceutical industry. The latter encompassed the International Federation of Pharmaceutical Manufacturers' Associations (IFPMA); individual companies such as GlaxoSmithKline, the largest beneficiary of the Mexican/swine flu pandemic; Becton, Dickinson and Co. (a medical equipment and testing company); Henry Schein (a dental equipment company that also sells Pfizer vaccines; this was the company that fired whistleblower Leonard Horowitz); and IBM (which had signed an agreement with Pfizer in late 2016 for developing immunological products by using artificial intelligence; notably a team led by the director of IBM Watson for South Asia had been the first to discover that SARS-2 was a laboratory product). Finally, there were also the medical faculties of two universities (Johns Hopkins and Georgetown), as well as Air Asia.

At this meeting, an important role was played by Edelman, the world's largest PR company. On its website, Edelman claims it introduced the concept of storytelling through the media (including the creation of the first 'media tour') to sell products and build corporate brands. 'With its foundation in consumer marketing, the company diversified to health, public affairs, technology and crisis management.' Edelman's annual Trust Barometer keeps track of the extent to which these 'stories' are actually believed—we already saw in Chapter 2 that this is going in the wrong direction for the ruling order. The term ruling class certainly applies to Richard Edelman, president and owner: he is a member of the Executive Committee of the Atlantic Council and has an impressive list of prestigious memberships including the CFR, WEF, Aspen Institute, Jerusalem Foundation, etc.[16]

In addition to Edelman, the BBC (itself a multi-million-dollar beneficiary of Gates Foundation largesse) was on hand to ensure that every story told about a pandemic would have maximum resonance and credibility. While without doubt the BBC has adhered to the script without fail, the major January 2021 study by Edelman made clear that this story was nonetheless losing credibility. The company termed this an 'information bankruptcy.'[17] But that lapse was not yet apparent at the WEF's anniversary meeting on the Spanish flu, which was in all respects a roll-call of the biopolitical complex.

A year later, in July 2019, a new transnational body, the Global Preparedness Monitoring Board (GPMB), released its first annual report. This organization had been convened by the World Bank Group and the WHO to prepare the world for 'the specter of a general public health emergency.' Specifically, it would be 'a very real threat of a rapidly developing, very deadly pandemic of a respiratory pathogen that would kill 50 to 80 million people and wipe out nearly 5 percent of the world economy.'[18] In reality, this Board was mainly aimed at preparing the ground for an international state of emergency, with strict obligations for individual states. The emphasis is to be on the 'prevention of dropouts; if one does not participate, all can perish.'[19]

Organizations broadly sharing this orientation had been set up before: the Global Health Crises Task Force and Panel was established by the UN Secretary General in the aftermath of the 2014–2016 Ebola epidemic. But if we need to identify one network in which the transfer of authority from legitimate government to a global governing body deriving its mandate from a biopolitical state of emergency, even an emergency not yet apparent, then it is the GPMB. This body would not decide what the new biopolitical disaster would serve to achieve (that would mainly fall to the World Economic Forum with 'The Great Reset'), but it would certainly determine its implementation, roughly according to the Rockefeller Lock Step scenario.

We can therefore safely leave aside the self-congratulatory praise of the 'independence' of the GPMB. Its only independence is from democratic control, parliamentary or otherwise. Under the dual leadership of Gro Harlem Brundtland, former Prime Minister of Norway and former WHO Director General, and Elhadj As Sy, Secretary General of the International Federation of Red Cross and Red Crescent Societies, the GPMB membership includes: the heads of the U.S. National Academy of Medicine, the Gates Foundation Global Development Program, the Wellcome Trust, NIAID (Fauci), and UNICEF, as well as the health ministers and similar officials of Chile, China, India, Japan, Russia, Rwanda and a few others. Financial support is provided by the German government, the Gates Foundation, the Wellcome Trust, and Resolve to Save Lives (a foundation funded by Gates together with Michael Bloomberg and Mark Zuckerberg). Working papers for the GPMB were provided by the Johns Hopkins University Center for Health Security and the University of Oxford in combination with the Royal Institute of International Affairs in London (Chatham House).

In its 2019 report, 'A World At Risk,' the GPMB instructs heads of government in each country to actually carry out 'their binding obligations under the 2005 International Health Regulations,' which we quoted earlier and which reflected the WHO's new authority. Such preparedness was to be understood as 'an integral part of *national and global security, universal health care and the Sustainable Development Goals* (SDGs).' It states that in a 'rapidly expanding pandemic of respiratory disease due to a deadly pathogen (whether naturally originated or released accidentally or intentionally),' donors and multilateral institutions would need to increase their investment in the development of innovative vaccines and therapies. Not a word about the general health of the population (which depends, to begin with, on security, adequate nutrition, water and air quality and proper living space, work and education, not on 'innovative vaccines and therapies'). But then, paradoxically, health is not a priority of the global biopolitical complex however purportedly mandated to protect global health. GAVI, the vaccine alliance, has an important role to play in preparation and supervision: the GPMB calls for 'new therapies and broad spectrum antivirals' and its objectives include (emphasis added): 'Distributed manufacturing of vaccines (including nucleic acid types) begins within days of obtaining the new sequences [of novel pathogens] and effective vaccines are pre-tested and *approved for use within weeks.*'[20]

The Johns Hopkins Center for Health Security too recommended 'gene therapy'—not incidentally a descriptor both Pfizer and Moderna would later apply to their products, possibly addressing the fact that these represented a departure from traditional vaccines[21]—in the aforementioned exercise Clade X, based on the scenario of a virus cloned in Germany and causing a global pandemic (In Chapter 3, I looked at the role of this exercise in the campaign to undermine Trump's reelection). Clade X was sponsored by the Open Philanthropy Foundation of Facebook cofounder Dustin Moskovitz, who also funds the Future of Humanity Institute at Oxford University. Nick Bostrom, the Swedish philosopher directing that institute, sees technological progress as fraught with danger, against which there is really only one remedy: mass government surveillance—in short, the scenario of the integrated surveillance society proposed by the intelligence-IT-media complex, one that must be internationally coordinated.[22] The Global Preparedness Monitoring Board too considers 'public safety protocols, school and business closures, aviation and transportation protocols, communication protocols, supply chain readiness' as measures to which states have already committed themselves in the 2005 agreements, in the wake of the SARS-1 scare.

Although the Board focuses on the tasks of the biopolitical complex, it is not blind to the political uncertainty accompanying the sudden introduction of the new scenario. It points out that confidence in institutions is eroding, as the Edelman barometer had already established several years earlier. Governments, scientists, the media, public health institutions and health professionals themselves are facing a collapse in public confidence in many countries. The situation is exacerbated by actual disinformation via social media, which can hinder disease control.[23]

Now if the EU was indeed viewed as the focal point in the concern of a looming multi-country popular uprising, with France as the hotspot, this was certainly borne out by plans to introduce a surveillance society under the guise of a pandemic. The European Commission had already published a proposal for vaccine passports in April 2018, and by early 2019, a roadmap to implement the options for a common biometric document. This is likely to be enshrined in binding legislation by 2022. In particular, 'innovative vaccines to deal with future health threats' had to be approved and the relevant industry given a key role.[24]

In September 2019 a 'vaccination summit' followed, organized by the European Commission together with the WHO, represented by its head, Dr Tedros. Again, the fight against disinformation concerning vaccines was high on the agenda. Political leaders and prominent personalities from the scientific, medical, business, and philanthropic worlds, as well as 'civil society,' had to be actively involved in this endeavor. However, the 'relevant industry' turned out not to be European, for the pharmaceutical companies represented here were Pfizer and Moderna (who would eventually produce the revolutionary mRNA vaccine) as well as Facebook.[25] Among the various round tables, the one entitled 'In Vaccines We Trust' was important because it compared the legal basis for vaccination in the different countries. Some EU member states recommend vaccination; in others it is required by law for everyone, or e.g. for school-going children, even kindergarten children. The U.S. with its comprehensive pro-vaccination regime served as the yardstick. However, the process of making vaccination compulsory, although already in some places very effective, had to be introduced with caution in order not to provoke anti-vax sentiment. Besides, compulsory vaccination is expensive and is only possible in high-income countries.[26]

For developing countries UN Secretary General Antonio Guterres and WEF head Klaus Schwab signed an agreement in New York in mid-June 2019 to speed up the introduction of the

Agenda 2030 for Sustainable Development. Roughly speaking, this was also the agenda that Schwab had laid down as the Great Reset in his eponymous book: the environmental and social pre-conditions for introducing the digital society into poor countries via public-private partnerships.[27] Eventually, once the 'pandemic' had become an accepted fact, 80 poor countries would be provided with IMF credits (with some debt forgiveness thrown in) of altogether one quarter of a trillion under certain conditions; key among them, the implementation of the Covid measures. The World Bank, partner of the WHO in the GPMB, had already pledged aid to the tune of $150 billion for 'affected countries' in February 2020. Particular attention was paid to India, Nigeria and Mongolia, although none of these countries at the time had identified a single case of contamination. Mongolia nevertheless introduced obligatory face masks and went into full lockdown. Nepal was also prepared for the worst. In April 2020, it received $26.9 million from the World Bank, followed by $214 million from the IMF in early May. This came after Nepal's imposition of lockdown in March, complete with quarantine camps.[28] The example of Belarus cited in Chapter 1, which declined an offer of nearly one billion from the World Bank given the conditions it had to meet, as well as the unexpected death of the President of Tanzania, thus can be placed in perspective, although the details in the latter case are still shrouded in mystery.

THE DRESS REHEARSAL: EVENT 201

In October 2019, in hindsight on the eve of the 'pandemic,' these by now familiar protagonists of the virus scenario—the Bloomberg-sponsored Johns Hopkins Center for Health Security (which had meanwhile come under the direction of one of the authors of the Dark Winter simulation of 2001), the Gates Foundation, and the World Economic Forum—convened a symposium in New York, titled Event 201—A Global Pandemic Exercise. This was a large-scale simulation, termed a 'germ game' (word play on 'war game'). The simulation of a virus outbreak (this time of Brazilian pig origin) was projected to lead to a death count of 65 million people over a period of 18 months, ushering in a world-wide economic depression that would last a decade, etc. etc. The participants included the Centers for Disease Control (CDC) of the United States; the head of the Chinese health authority, U.S.-trained Dr. George Gao; Avril Haines, a former deputy director of the CIA, now a lobbyist for IT companies with the U.S. government and the intelligence services; as well as representatives of the pharmaceutical sector, notably Johnson & Johnson and Henry Scheiu (agent for Pfizer vaccines), as well as the health and risk specialists of several other large companies.[29]

Event 201 once again put great emphasis on the danger of an 'infodemic,' a tidal wave of 'disinformation' questioning the mainstream narrative. One of its recommendations concerned tightening the censorship of social media and even closure of the Internet altogether. In part four of the script, the tone was set by Matthew Harrington, CEO of Edelman, the PR concern that had taken care of the storyline for a pandemic scenario in previous exercises. Harrington pointed out that the IT giants who run the social media platforms needed to recognize that in a pandemic they should move into the role of broadcasters, that is, selecting news in line with a clear editorial policy. Their task was to 'flood the zone' with the right viewpoints. Therefore, he suggested setting up a central hub from which the correct perspective and 'key messages' could be transferred to lower levels.

Other participants supported Harrington's call for tighter control over social media, but Chen Huang (Apple/Google) went one step further and argued that the Internet should be shut down if necessary 'to avoid panic.' Admiral Stephen Redd of the Public Health Service even came up with the idea of tracking down people with 'negative beliefs' on social media; a huge job but, as another speaker added, if it wasn't taken on, governments might begin to fall, as had happened in the Arab Spring. Singapore's finance minister even advocated arresting dissidents and bring them to justice.[30] This reflects the original stimulus that gave rise to the pandemic state of emergency scenarios: the fear that the world's population can no longer be contained.

In addition, the particular interests of the pharmaceutical companies were also considered. Dr Gao (who is also a member is the Global Preparedness Management Board) warned that there might be 'rumors' that the virus originated in a laboratory and that a pharmaceutical company was behind it. Johnson & Johnson's Adrian Thomas in turn expressed concern about a possible public perception that poorly tested vaccines could have fatal consequences and wondered if his company and the competitors would be doing the right thing by reporting deaths and infections in trials.[31] How the participants could have had such an accurate sense of what the 'disinformation' would focus on was not only remarkable, but reflected their awareness of the already existing high level of public distrust—and indeed its perception of their conflicts of interest.

This brings us to the 'vaccines,' in reality, as yet untested so-called gene therapies, to which governments, parliaments and the mainstream media have committed themselves in advance and without reservation and from which they tolerate no dissent.

THE BRAVE NEW WORLD OF mRNA VACCINES

Vaccination has made an important contribution to the health situation in rich countries with high sanitary standards, although it has often been pointed out that it poses specific dangers even there. Nevertheless, vaccination has gradually risen from a supporting facility to a paramount instrument of health care. In the current Covid crisis, this has reached the stage where all precautions have apparently been thrown overboard. Vaccinating more than 7 billion people who are mostly not ill is an Eldorado that a whole host of pharmaceutical explorers have set out to discover. After all, 'cure' is not really a revenue model. It is much more profitable to make people dependent on drugs irrespective of whether they are in danger of falling ill in the first place, and that precisely is what is made possible by the current crisis. Indeed, the administration of intravenous gene therapies that produce unexpected effects can even initiate, intentionally or unintentionally, a permanent follow-up treatment regime by way of boosters, or yet additional therapies to deal with virus mutations possibly caused by the vaccines. After all, the pandemic may primarily be a pandemic of fear, but there is a real chance that applying rushed medication on a grand scale will cause a real disaster.

Serious diseases can already be caused by growing viruses for vaccines on animal tissue: from autism and chronic fatigue to cancer or Alzheimer's. In 40 percent of patients with early prostate cancer, integrated sequences of a mouse retroviral virus were found in the tumors. In 2012, Harvard researchers reported to the WHO that xeno-transplantation from species to species might introduce foreign pathogens into the human population.[32]

Nevertheless, the philosophy underlying Western medical science, like it does so many aspects of capitalist society, has increasingly shifted towards technical-scientific remedies. Within it, chemical and biochemical solutions have in turn taken precedence. The medicalization of social and psychological problems and attendant profit opportunities for the biopolitical complex, in particular the pharmaceutical industry, promote this view at the expense of promoting and funding innate immunity, a healthy lifestyle, and the like, about which not a word is heard in the current Covid state of emergency.

FROM mRNA TO MODERNA

We saw in Chapter 4 that the process by which viruses enter the body can be compared to 'updates' of a computer operating program. With such an insight, it was then only a matter of time before the idea will arise that what nature can do, man can do especially if incentivized by commercial application. In 2010, Canadian scientist Dr. Derek Rossi, working at Harvard, found a way to 'reprogram a cell.'[33] This 'messenger ribonucleic acid' (mRNA) contains the necessary instructions to influence ('express') the production of protein in the cell. At the time, many researchers spoke highly of the future therapeutic manipulation of viral material in humans,[34] despite the fact that disruptions of the microbiome are already associated with non-communicable diseases and autoimmune disorders that are increasingly common in the industrialized world.[35]

When Rossi retired in 2014, he teamed up with investor Flagship Pioneering to monetize his discovery, Moderna (a contraction of modern and RNA). In 2016, Flagship Pioneering issued a press statement highlighting parallel collaboration with AstraZeneca, Bayer Crop Science and Nestlé Health Science,[36] corporations already engaged in the genetic engineering of food crops and products. Moderna's website provides an insight into the vision of the creators of this new technology. According to the company, the mRNA platform should be seen as the operating system of a computer, where the mRNA drug functions as a program, or 'app.' What is revolutionary is the genetic code that instructs a ribosome to make a protein. The mRNA provides the software and can be combined at will (i.e., different mRNA codes for different diseases). However, the functioning of the immune system must be circumvented because otherwise it will not work; the ribosomes must respond as if it were a natural mRNA and read the code precisely.[37] Now that is just what the engineered SARS-2 virus does as well, it enters the cell without triggering an immune response.[38]

Both Moderna and Pfizer (along with Germany's BioNTech) would later come up with an mRNA vaccine that exactly mimics the HIV spike protein discovered by Pradhan's group to activate the immune system from within.[39] Initially, the goal of both Moderna and BioNTech had been to develop a cancer therapy, but the SARS-2 alarm cut across that route. With unprecedented profit opportunities in the offing, they promptly switched to combating the 'pandemic.' Meanwhile, Moderna, in which Rossi himself is no longer active, has secured half a billion dollar in orders from the U.S. federal government thanks to its connections to both

CEPI (one of the branches of the Gates Foundation) and DARPA, the research arm of the Pentagon.[40] With the mRNA technique, the aim is now to vaccinate as many as possible, preferably the entire world population, with a greatly shortened test period in which the phase of animal experiments is skipped entirely.

The idea is that the protein of the virus to which the immune system has to respond is no longer made in the laboratory, but in the human cells themselves, prompted by the mRNA introduced from the outside. The immune system is then ready, when a similar pathogen comes along, to fight it off. However, previous animal experiments (not for Covid, where animal experiments have been skipped) have led either to immediate death, or in the majority of cases, to death after another virus was contracted later. If this new virus is related to that encoded in the injected mRNA, there can occur an over-response (cytokine storm) in which the immune system attacks one's own body, with sepsis and organ failure as a result.[41] Indeed, any virus entering the cells can now activate a super-antigen encoded by an endogenous retrovirus. Such a super-antigen, a bacterial or viral poison, causes an immunological overreaction, read: autoimmune disease. Because the cell's response systems are not always able to distinguish between human and viral DNA and RNA, especially when cells are overwhelmed by their residues, this can have extremely grave autoimmune consequences, such as multiple sclerosis and Alzheimer's.[42]

Professor Dolores Cahill of University College Dublin was one of the first to highlight these dangers. Upon declaring that Covid-19 was a cover to deprive the population of its rights and submit to forced vaccination, she was removed from her position as vice-chair of the European Committee for Innovative Medicines. Cahill warned that in addition to the autoimmune diseases over time, death could also occur if another virus were to enter the body within a few months or a year following inoculation.[43] Several previous studies had pointed in the same direction. Following the SARS outbreak of 2002–03, a vaccine had already been sought and DNA therapy had also been tried on mice; the vaccines were shown to be effective against SARS but also caused immune disorders in the lungs, indicating hypersensitivity to components of the SARS virus. At the time, researchers warned against applying this to people insofar as it is always risky to vaccinate against one respiratory virus because it can increase the susceptibility to others ('virus interference').[44]

Not all pharmaceutical companies follow the Moderna and Pfizer/BioNTech mRNA pathway. AstraZeneca and Johnson & Johnson and its Dutch subsidiary, Janssen, developed a gene therapy with which genes from SARS-2 can be introduced into the cell via another virus (in both cases, the question arises as to where they got those 'genes from SARS-2,' because whether this virus has been isolated at all is still in doubt). Novavax and Sanofi Pasteur use a SARS-2 protein or protein fragment to elicit the immune response; the Russian Sputnik-V follows the approach of AstraZeneca which would later seek the Russian company's help for its own vaccine. Only Chinese Sinovac uses a weakened SARS-2 virus as an immune stimulant, the usual route to obtain a vaccine. So this is the only substance that can claim the title 'vaccine' and falls under the appropriate legislation.[45]

Parallel to the aforementioned exercises and simulations, there were also preparations for a large-scale deployment of these revolutionary gene therapies. As early as June 2019, again before anyone knew about a possible new virus, Dietmar Hopp's German biotech company CureVac, together with Tesla subsidiary Grohmann (owner, Elon Musk), registered a patent on RNA vaccine 'printing plants' which will be able to produce them in record numbers.[46] The Gates Foundation had taken a $2 million share in CureVac as early as February 2015 as the company was developing an mRNA vaccine; after that, an additional $2 billion was pledged. The German government in turn acquired a share of 300 million euros to keep it in German hands.[47]

Another serum dispenser comes from Inovio, a competitor of Moderna specializing in genetically engineered immunotherapy and vaccines using DNA plasmids. Inovio is funded by USAMRIDD at Fort Detrick to develop a portable device for delivering these inoculations. The company has experience in genetically engineered vaccines for coronavirus infections such as MERS, and has secured a CEPI grant to develop a Covid-19 vaccine. The MERS inoculation is being tested in the Middle East at the time of writing.[48]

DARPA and the aforementioned Defense Threat Reduction Agency (DTRA) have also funded Inovio to develop an Ebola vaccine. In addition, the company has a genetically engineered vaccine for the Zika virus, but it was not approved for use in humans in the United States. Outside the U.S., there is a lot of experimentation with vaccines: two major programs started in Georgia in 2013: one, under the auspices of NATO (!), investigated anthrax vaccinations in humans (the previous program for anthrax vaccinations in livestock was closed in 2007); the other was a DTRA project on Tularemia disease, which lasted until 2016.[49]

It appears again and again that behind both the biopolitical and the IT-media power blocs lies the strategy of the American national security complex. In January 2020, Luciana Borio, of the National Security Council (where she was responsible for bio-defense) and meanwhile vice-president of In-Q-Tel, the CIA's investment arm, declared that in the Covid crisis, then in its early stages, the development and use of new diagnostic and subsequently therapeutic instruments had to be made the key concern.[50]

One of the most alarming aspects of the transition from mechanical to psychological warfare against the population is that the authorities have now set their sights on the human genetic code as well.[51] The express encouragement (also by the Global Preparedness Monitoring Board) of nucleic acid-based vaccines makes clear that this is a deliberate strategy. The slogan '*leave no one behind*,' a slogan that was never used when it came to combating malaria, other diseases, poverty, or even the world food situation, speaks volumes here.

THE VACCINE BONANZA

Following the official proclamation of the 'pandemic,' the major pharmaceutical companies plunged into the market for vaccines, mostly in conjunction with academic or private biotech start-ups. They were obviously well aware that the politics of fear creates unprecedented profit opportunities for them. Vaccination of the entire world population of 7 billion, soon 8 billion people, and preferably, made compulsory, represents a market that has existed never before. With few exceptions they placed their bets on vaccines aimed at smuggling genetic instructions into cells to get these to make the spike proteins which the immune system then responds to as if they belonged to the coronavirus. The three main vectors used were lipid nanoparticles with mRNA (Pfizer/BioNTech, Moderna, CureVac); genetically modified adenoviruses (AstraZeneca, Johnson & Johnson/Janssen), and DNA plasmids (Inovio).[52]

There is one problem imperiling the pharmaceutical companies' bonanza, and that is that vaccines may not offer the only solution. Hydroxychloroquine (HCQ), the anti-malarial drug, has already proved a successful first-line treatment for Covid. This has been freely available since 1955 and in combination with zinc and an antibiotic (azithromycin), it was also found to be effective within four days for patients who reported symptoms early. Positive results with HCQ were reported in China in mid-February 2020; in the U.S., New York doctor Vladimir Zelenko came up with the three-component protocol, and academic specialists also recommended the drug.[53] In France it was recommended (always in the aforementioned combination and in the first stage, not later) by Professor Didier Raoult of IHU-Mediterranée in Marseille, one of the world's top five scientists in the field of communicable diseases. Raoult came under fire from the Macron government and Sanofi-Pasteur, which itself was in the running for producing a vaccine. In January 2020 HCQ was removed from the list of over-the-counter medications by the French Minister of Health, the wife of a top executive of the Institut Pasteur; henceforth a prescription was required. This was a possible indication that the pharmaceutical interests saw HCQ as a competing, and hence profit-diluting, remedy.[54] Next the French Infectious Diseases Society Spilf filed a complaint against Raoult in the autumn of 2020 for allegedly promoting a 'new' treatment method to the general public, the effectiveness of which had been insufficiently demonstrated.[55] According to the *Hold Up* documentary, Raoult also received a death threat, as it turned out from a colleague with close ties to the American pharmaceutical concern Gilead Sciences, and in June 2021, police even entered his institute in Marseille on a suspicion of corruption.[56] Elsewhere, similar measures to suppress HCQ as a possible treatment were taken: in the Netherlands, the public health authority bought up the stock of HCQ at the end of March 2020 and a first-line doctor was instructed to stop prescribing it in spite of obvious effectiveness; in Britain, an export ban of HCQ was issued. NIAID's Fauci also spoke out against HCQ, possibly because, as a financial expert quoted by Sandra Depuydt candidly observed, if a Covid drug doesn't benefit a stock, it doesn't exist.[57]

Subsequently, other simple and effective agents came into the picture as first-line treatment such as Ivermectin, but it was now clear that considerations other than medical ones were involved here. If anything, Facebook, LinkedIn and Twitter as well as the Google subsidiary, YouTube, made no secret of their intent to follow the official line that these cheap medications were to be made suspect in the eyes of the public.[58]

The pharmaceutical industry had meanwhile entered the race for billions' worth of vaccine orders and, as with the Mexican/swine flu crisis, it once again sought to exploit an alleged 'emergency' to accelerate the search for new therapies, this time experimental gene therapies—this time with the advantage of huge preorders and protection from liability. Since a provisional approval from the supervisory authorities can only be granted if in an emergency there is no available alternative, one can begin to understand the bans on HCQ and Ivermectin as a precondition for marketing hastily developed medication by the pharmaceutical giants.

AN INDUSTRY WITH A CRIMINAL RECORD

The pharmaceutical industry has a record that would put a top criminal to shame. At the time of the Mexican/swine flu of 2009, it was reported that one-tenth to a quarter of the turnover of the pharmaceutical industry disappeared due to corruption (fraud and bribery); this criminal behavior is the rule and the sector continues to be inundated by fines, which usually end with a settlement but nevertheless indicate the seriousness of the crimes committed. This has not changed. The fraud involved here includes all forms of deception, forgery, and lying about the safety of medications. In 2003, *The New York Times* revealed that in the 1980s, Bayer had supplied millions of hemophilia patients with a blood coagulant that was contaminated with the HIV virus in the case of Asia and Latin America, whilst delivering a safe version to patients in the West. Ultimately, Bayer was forced to settle for $600 million.[59] Between 1991 and 2015, 373 settlements in all were reached between the U.S. authorities (mostly the federal government) and the pharmaceutical industry, for a total of 35.7 billion dollars.[60] In an update for 2016–17, *Public Citizen* released figures indicating that the downward trend in fines had not continued in 2014–15, as federal settlements totaling $2.8 billion were reached again despite lower average fines.[61] In addition, the U.S. Institute for Vaccination Side Effects Compensation (VICF) has paid out more than $4 billion in damages since 1988, of which more than 3.7 billion since 2006.[62]

As the Bayer example shows, the European pharmaceutical sector is also regularly caught in all kinds of fraud and deception. As late as 2020, the French competition authority fined the pharmaceutical groups Novartis, Genentech (a U.S. firm) and Roche for a combined amount of $444 million for fraud.[63] Roche among other things produced Tamiflu, claimed to be a cure for the 2009 Mexican/swine flu; the *British Medical Journal* later revealed that the experts who advised the WHO to use this and other antivirals had secret contracts with manufacturers. In 2017, the WHO removed Tamiflu from the list of essential medicines because it doesn't work. The laudatory articles that had appeared recommending it appeared to have come from commercial copywriters who worked under the direction of the manufacturer, Roche. In the U.S., Roche was fined $59.7 million in 2020, but the company is nonetheless back in full force for a Covid treatment.[64] Sanofi–Pasteur, which has marketed an anti-epileptic drug since 1967 and which it knew from the early 1980s that it might lead to malformations in pregnancy, was only indicted in 2020; the French state was meanwhile sentenced to pay damages to victims. Sanofi-Pasteur is in the race for a Covid vaccine with two projects with different biotech partners, one based on a medication for rheumatism, the other a nucleic acid vaccine; the fact that it was fined $152 million in the U.S. in 2012 and 2013 apparently did not disqualify it for entering the market with an experimental treatment. Another French company, Servier, from 1978 marketed a treatment for diabetes (Mediator) that was not only ineffective but also toxic, with heart problems as a result. Despite this, the expensive drug remained on the market until 2009. In March 2021 Servier was found guilty but appealed.[65]

In addition to the actual pharmaceutical companies, there are also distributors who have made a name for themselves in fraud cases. First Databank (FDB), a subsidiary of the Hearst media group, which sells vaccines online, and the aforementioned Henry Schein, which sells Pfizer vaccines in addition to its own dental equipment, were both involved in fraud cases.[66]

One of the practices of the pharmaceutical industry in developing new medications, once the animal trials (which were skipped in the race for a Covid vaccine) have been completed, is to undertake experiments on human subjects in poor countries. Until recently, Africa was the designated continent for this. In the years 1997 to 2003, HIV-1 inhibitor Nevirapine was tested on Ugandan women, who experienced serious side effects and some of whom died. As there was no report on this, testing continued. India is also a popular field of this sort of activity. In 2003, eight people in Hyderabad (unaware that they were participating in an experiment) died after testing for Streptokinase; in total, according to Indian government data, nearly 5,000 people died from testing and research between January 2005 and November 2017.[67] In addition, it is being suspected that vaccination is used for birth control. In Kenya in 2014, the WHO and UNICEF administered a tetanus injection in an unusual way (five doses) and only to women between 14 and 49 years old.[68] In July 2020 there were protests in South Africa against the testing of Covid-19 medication on poor, uneducated people whilst cheap and widely available drugs and traditional medicine were being suppressed.[69]

Given the track record of the companies currently competing to get their Covid medication on the market or still developing it, one might well expect that safety and product liability issues would be paramount in negotiations. While the EU actually did place product liability on the agenda although its subsequent bulk purchase of Pfizer vaccines (see below) did not contain clauses covering that as far as we know, in the U.S., pharmaceutical companies are covered by government guarantee under the National Childhood Vaccine Injury Act of 1986, which drastically limits normal company liability for damages.[70]

Finally, mention should be made of the corrupting influence that pharmaceutical companies have on medical journals, which provide scientific support (or not) for their business strategies. It turns out that under pressure from industry, about half of clinical trials are ultimately not published, usually because they reported on negative outcomes.[71] In apparent contrast, an article labeling HCQ as dangerous was published in the renowned *Lancet,* and widely cited in the mainstream media. It had to be withdrawn again in June 2020 after it turned out to have been based on fraudulent data, although this retraction did not receive much media attention. When GlaxoSmnithKline (GKS) introduced an antidepressant in 2001, a study was published that declared the drug effective and risk-free; only in 2015 did the *British Medical Journal* reveal that a risk of suicide had also been identified. The 2001 study turned out to have been written by a ghost author on behalf of GKS. In 1999 Merck, the world's oldest pharmaceutical company, released an anti-inflammatory drug, Vioxx, which led to cardiac arrest in tens of thousands and was only withdrawn from the market in 2004. Authors employed by Merck had written that the drug was well tolerated, although tests had shown a high mortality rate.[72] Then as mentioned, there were the ghost writers who in the pay of Roche lavished praise on Tamiflu.

The editors of medical journals would like to see themselves as gatekeepers for accurate information, but the pharmaceutical industry is the co-author of everything they publish and the revenue associated with reprints of articles, purchased by the company whose products are vindicated in them, is a vital source of income (in *The Lancet*, more than half of all revenues). Clearly this works to make corruption difficult to avoid. According to the former editor of *The New England Journal of Medicine*, who had 20 years of experience, it is simply no longer possible to believe much of the published clinical research.[73]

FRONT-RUNNERS IN THE RACE FOR A GENE THERAPY

Which companies are now playing the leading role in the 'fight against Covid-19'? I put this in quotation marks because 98 to 99 percent of humanity does not even notice that they are infected with this virus plus that it would in principle be mutating from highly virulent to more infectious but less dangerous variants anyway. 'In principle' because in this case, there may be unexpected consequences due to the inoculation with experimental vaccines that may make the vaccinated more vulnerable to a new iteration. Let us go over the main beneficiaries of the Covid panic and the readiness of governments to propagate the substances sold as vaccines. Russian, Chinese and Cuban alternatives are left out of the picture, because my focus here is not on public health but on disaster capitalism and profit opportunities, a phenomenon that has its basis in the West.

AstraZeneca. The company which originally appeared in the lead for a revolutionary vaccine against Covid-19 was Swedish-British AstraZeneca. It had entered the competition with Oxford

University as its scientific partner. AstraZeneca was never a significant vaccine manufacturer, but it ranked 10th among the most fined pharmaceutical companies in the world, incurring a total of $1 billion in fines in the U.S. alone. The largest fine (ultimately a settlement) dates back to 2010 and concerned deception regarding the safety of one of its drugs.[74] Why would any government ever select a company like this for such a sensitive, possibly very risky product as an experimental gene therapy? I will take the Netherlands as my example because of the spectacular corruption accompanying the selection of AstraZeneca in this country, as I show at length in the Dutch version of this book. Here I summarize.

In March 2020, the Dutch health minister appointed the outgoing chairman of the board of the country's chemical and biotech giant, DSM (who had resigned one month earlier), to the new post of 'corona envoy' with the remit to fix the shortage of tests and find a vaccine.[75] Soon DSM signed for the supply of 2.8 million swabs for PCR tests, but the corona envoy also brought in AstraZeneca, the company where his own brother made his career, eventually ending up as president of AstraZeneca Germany. Thereupon the health ministry signed a contract for 200 million euros, reserving another 700 million for further purchases. This contract was declared a *state secret*, according to the health minister because information about it might 'cause serious damage to the interests of the Netherlands or its allies.' What 'allies' have to do with the 'pandemic,' a real parliament would have wanted to know, but the Dutch lower house by then had already resigned its role as a representative assembly serving the interests of its public, as had so many parliaments in other countries. It was only after a private citizen asked the Dutch prime minister in August about the family connection behind the choice for AstraZeneca that the corona envoy announced he was stepping down.[76]

This was the beginning of the undoing of the one European front-runner in the vaccine contest. In its contracts with governments, AstraZeneca stipulated, even before any vaccine had been developed, that the company would have immunity from liability for side effects for four years. According to its CEO, this was 'a unique situation where we as a company simply cannot take the risk,' just as it was up to AstraZeneca to judge when the 'pandemic' would end because then the grace period in which it would supply the vaccine at a reduced price would end as well (the expiry date was originally set for July 2021, but the company may then decide 'in good faith' that the time has not yet come).[77] AstraZeneca had to come up with something at record pace (the race was eventually won by Pfizer, see below). In December 2020, the company announced that the 'vaccine' developed with Oxford was 'safe' and (in a second press release) that the desired immune response had occurred in the elderly age group. The ChAdOx1 nCoV-2019 vaccine was based on a genetically modified adenovirus (a flu-like virus that can infect the upper respiratory tract) in combination with the spike protein of SARS-2 to bypass the immune system and enter the human cell.[78] The danger with this type of vaccine is not only that the adenovirus can recombine with natural viruses and form hybrids with potentially very dangerous properties; the spike protein, too, has turned out to be a dangerous molecule because of its clotting properties. As we will see below, these clots can become manifest in any part of the body because, unlike normal vaccines, the spike protein does not remain in the injected area, e.g. the upper arm, but circulates through the bloodstream.

Meanwhile, the EU too had signed a contract with AstraZeneca (the first EU contract) in August 2020, which included a partial liability cover. The European Commission, like national governments, wants to keep alive an atmosphere of fear in order to cut off debate and pave the way for directly or indirectly mandatory vaccination. The genetic modification of plants and animals is subject to strict rules in the EU, let alone that people could be subjected to treatments relying on the introduction into the human body of viruses that have been genetically modified. According to European Commissioner Kyriakides, there was no time to lose. 'As long as there is no vaccine, no one is safe,' she declared, temporarily putting aside the rules for new drug testing procedures. In November 2020, a pharmaceutical industry lobbyist would take office as head of the European Medicines Authority EMA and grant the provisional approval.[79]

In the meantime, AstraZeneca was testing the vaccines on humans in the U.S., Brazil and South Africa, and from August also in India (there was no time for animal testing); Oxford tested it in England. Tests were discontinued in September 2020 because a British participant, a 37-year-old healthy woman, developed transverse myelitis caused by inflammation of the spine several weeks after taking the second dose. AstraZeneca issued the standard statement that the condition was unlikely to be vaccine-related, etc. and resumed testing in Brazil and Great Britain. In the U.S., on the other hand, it was not allowed to do so before safety concerns were met.

After initially ranking highest on orders, the AstraZeneca star began to fade when, shortly after, the parent company and Oxford University contradicted each other on the necessary dose. Subsequently, AstraZeneca announced that it would collaborate with Sputnik in Russia, to see whether the effectiveness of its own vaccine could be improved by including Sputnik's V vaccine, which is also an adenovirus vector vaccine. The company then hit the headlines with a planned $39 billion mega acquisition of U.S. pharmaceutical group Alexion (to give an idea of the competition among pharmaceutical companies, AstraZeneca itself had fended off a hostile $86 billion acquisition by Pfizer). All in all, these moves contributed to a sharp fall in the price of AstraZeneca's share. Because of these issues, notably bad publicity regarding the blood-clotting effect, AstraZeneca has ended up last in the race and all hopes of a 'European' solution vanished accordingly.[80]

Before moving to the other companies, let me briefly note how the actual testing on the basis of which the provisional approval was granted, has been evaluated. William Haseltine, a former professor at Harvard Medical School, found that only young and healthy subjects were injected in all tests. As a result, the numbers paraded (for AstraZeneca, 18,000) were impressive but in fact, test results were based on less than 100 cases. At AstraZeneca, the test protocol was that if out of fifty test persons vaccinated, less than twelve developed symptoms, whilst of the control group of 25 who were not vaccinated, 19 people had symptoms, the vaccine was considered effective.[81] Below we will see that the other top contestants based their results on test protocols that were equally minimal. The AstraZeneca test protocol (and the same holds for the test protocols of Pfizer, Moderna and Johnson & Johnson) did not address whether the 'vaccine' actually prevents the virus from spreading. Of course, today, with mass vaccination in full swing, the 'testing' has in fact exploded as well, but the figures cited were what the provisional license was based on, and on which millions of people based their confidence that they were in safe hands.

Pfizer. Among the large pharmaceutical companies, the long-established U.S. giant, Pfizer, and the newcomer, Moderna, apply the mRNA route. Pfizer worked with a German biotechnology firm, BioNTech, set up by two Turkish-German microbiologists who have since become billionaires themselves. Pfizer/BioNTech actually was the first combination to announce that its quest for a vaccine had been successful. Like Moderna's, its product also contains nanoparticles, to which I will come back separately below. Tests on human subjects (here too, animal experiments were skipped) began in late July, early August; on 9 November, the good news was made public. Naturally, the company's stock market value climbed, and Albert Bourla, Pfizer's CEO, sold two-thirds of his shareholding in the company on the day of the new product's disclosure, securing millions in profit just in case matters went south. (So as not to be accused of insider trading, he had already taken out an option for such a sale in August).[82] In August 2019, Pfizer set up a joint venture with GlaxoSmithKline, the most successful disaster capitalist in the Mexican/swine flu crisis. The joint venture was meant to create, in the words of a GSK press release of 1 August, a new, world-embracing pharmaceutical and vaccine group that focuses on knowledge about the immune system, the use of genetics, and new technologies.[83]

The fact that Pfizer, after hasty testing, made its success public through press releases (instead of scientific publications and their time-consuming peer reviews) becomes even more problematic when we review the criminal record of this company which meanwhile has become the winner of the corona-vaccine race. Over the period 2000–2015 and in the U.S. alone, Pfizer chalked up a total of $3.4 billion in fines (through settlements). The fines were handed out in 2004, '09, '13 and '15, so we are looking at a repeat offender. 2009 was the highest ever fine for medical fraud ($2.3 billion).[84] The U.S. was not its only arena of misconduct. In the same year 2009, the company announced it had paid out $53 million in delayed lawsuits over child death liability in Nigeria over the bacterial meningitis drug Trovan it had administered there in the late 1990s.[85] In 2004, Pfizer's interest in new vaccines and medications led it to sign an agreement to work on a new generation of drugs with Genstruct, a company funded, like Moderna, by Flagship Pioneering. In 2014, Pfizer began funding Johns Hopkins University to undertake a research program on 'Strategic Immunization in Low and Middle Income Countries.'[86]

Not unexpectedly Dr Anthony Fauci was one of those who touted Pfizer's new drug as a milestone and predicted that it would soon be approved. The risks mentioned earlier were left aside, as were the possible consequences of the introduction of nanoparticles into the human body. Fauci himself is actually a co-patent holder of HIV-related peptides used in the gene therapies of both Pfizer and Moderna. After the U.S. Department of Health and the Pentagon placed an order of 100 million doses for $195 billion with Pfizer in July 2020, the month that trials began, with the option for a further 500 million doses (like Moderna's and AstraZeneca's, the Pfizer vaccine requires the administration of two doses), approval was a foregone conclusion.[87] Not to be left behind, the EU announced that it had also purchased 300 million doses of the precious vaccine and in May 2021, after the demise of AstraZeneca's product, even followed up with an order of 1.8 billion doses of Pfizer/BioNTech, four for every EU citizen, man, woman and child.

Moderna too shared in the bonanza. Following the announcement in April 2020 of a U.S. government subsidy of just under half a billion, its stock skyrocketed. Like Pfizer's Bourla,

Moderna's executives did not fail to cash in. In the course of 2020, CEO Stéphane Bancel sold $40 million in Moderna shares, Chief Medical Officer Tal Zaks sold around $60 million and President Stephen Hoge, more than $10 million, all in this instance raising the insider trading issue.[88] From 1 May 2020, the company also became the 'strategic partner' of the Swiss pharmaceutical group Lonza, which would produce a Moderna mRNA vaccine under license. A few days earlier, Moderna director and Glaxo veteran Moncef Slaoui had been appointed to the board of Lonza; on May 15, he was appointed by Trump as project leader of Operation Warp Speed, to accelerate the search for a vaccine with an eye to the president's reelection. Slaoui stepped down as director of Moderna but was allowed to keep his shares (his share options alone had risen nearly 200 percent in value).[89]

In late 2020, Pfizer/BioNTech announced that the mRNA-based BNT162b2 'vaccine' had achieved a 90 percent effectiveness score; Moderna's mRNA-1273 topped that at 94.5 percent. BioNTech also shared the BNT162b1 serum with Shanghai Fosun Pharmaceutical Co., Ltd (Fosun Pharma) to develop a version for the Chinese market.[90] However, these apparent PR successes were not taken for granted by the scientific community. The *British Medical Journal* pointed out that the two companies had published incomplete results, whilst the long-term side effects are unknown, and the risk of contracting Covid-19 is only slightly reduced (by less than one percent).[91] The previously quoted Harvard emeritus, William Haseltine, also disqualified the Pfizer and Moderna test protocols. Here too, large numbers appeared in the companies' PR but in fact, the results were based on fewer than 100 tests. Pfizer even had the smallest group of all, 32 vaccine recipients; at seven or fewer with symptoms its vaccine was considered successful (with a control group of 25). At Moderna, 53 people were injected; if there were 13 or fewer with symptoms (as compared to a control group of 40), the product was seen as a success.[92] This was the basis upon which millions of healthy people are meanwhile being injected with a revolutionary drug that in its final form has been around for less than a year or two.

Moreover, as we will see further on, this vaccination campaign has the potential for triggering a real pandemic.

The *BMJ* also published a letter from a group of professors from four countries commenting on Pfizer and Moderna (as well as AstraZeneca) press statements. The authors asked why these results were not broken down by age category; why it had not been reported how they affected the transmission of the virus and the length of time the protection would be effective. Finally, they asked why all of this had not been published, as usual, in scientific journals instead of via press releases. The authors considered this as compromising the integrity of the testing process and undermining public confidence in the vaccines. It need not surprise that the Pfizer 'vaccine' test was paid for by Pfizer.[93] Indeed the head of the infectious diseases department at the Pitié-Salpêtrière hospital in Paris warned that launching these new therapies with their unforeseen consequences in this way ran the risk of discrediting vaccination as such.[94]

The complainant in the aforementioned Florida trial, the whistleblower Dr. Leonard Horowitz, had also studied the Pfizer and Moderna protocols and came to the same conclusions. In addition, the presence of an HIV gene in the Covid-19 vaccine had been concealed. Horowitz had written a book about his experiences (he had been fired by Henry Schein in the early 1990s after revealing that this firm was responsible for infecting a number of dental patients with HIV) and in 2008 launched his own anti-microbial and anti-viral drug.[95] The trials, Horowitz said, were inconclusive about the consequences of using the spike protein that mimics HIV and increases the transmissibility of the virus. It can therefore cause new, unknown infections. Pfizer even stated that genetics has not been taken into account; it had excluded from the trials virtually all risk categories for which the supposed protection is actually needed. No wonder Dr Tedros, the head of the WHO, issued a statement that even after the administration of the 'vaccine' the virus could still spread—as we see below, possibly even in a more dangerous form.[96]

Dr. Michael Yeadon, who has worked in pharmaceutical research for three decades and in one of his roles was vice-president for research at Pfizer, declared that if an mRNA vaccine would be approved without the express clarification that it might be used only for experimental applications, this would amount to criminal deception.[97] As we saw however, Pfizer is no stranger to criminal deception, on the contrary. Below I come back to Pfizer because of a unique and very disturbing experiment being conducted in Israel, where the company has been granted a monopoly and the government of Benjamin Netanyahu (whose own trial for fraud began in March 2021 and who has meanwhile been replaced) is putting the population under severe pressure to be injected.

Johnson & Johnson. In the Public Citizen overview of companies' criminal records, Johnson & Johnson is in 3rd place with total fines of $2.8 billion for the period 1991–2015. This has since risen sharply: in 2018, the company was fined $4.7 billion for its cancer-causing baby powder.[98] It is also involved in the opioid scandal in the United States. Its use as an (addictive) pain killer has been identified as one of the causes of the decline in life expectancy in the U.S. Along with Fentanyl, it has caused hundreds of thousands of deaths due to overdose, although the prime manufacturer, Purdue Pharma, continued to urge doctors to prescribe it. Johnson & Johnson has also received claims for this and will have to pay more than $8 billion in fines and damages once it pleads guilty.[99] All this has hardly worked to disqualify Johnson & Johnson from entering a new and risky field; but then, the related Robert Wood Johnson Foundation is the largest donor to medical studies in the U.S.[100]

The Johnson & Johnson vaccine is based on a single dose of a genetically modified cold adenovirus that cannot reproduce, combined with the spike protein of SARS-2 (Ad26.COV2.S) that allows it to enter human cells. The company has received $1.45 billion from the federal government as part of Warp Speed. In October 2020, however, Johnson & Johnson announced that it had stopped testing its Covid-19 vaccine because one of the participants had become inexplicably ill. Subsequently, its opportunity to register for one of the 60,000 available test slots (distributed to over 200 locations in the U.S., Argentina, Brazil, Chile, Colombia, Peru, Mexico, and South Africa) was suspended pending a ruling of the committee that investigates 'serious adverse events.'[101]

This registration for '60,000 test slots and 200 locations' again sounds like a lot, but in the Haseltine comparison of test protocols this does not appear warranted. In the case of Johnson & Johnson, 77 people infected with SARS-2 were treated with vaccine; if fewer than 28 of them acquired symptoms, the vaccine was considered a success. That we are looking at a heavy drug was clear from Johnson & Johnson's warning that fever, muscle pain and headaches can occur and that these can be particularly severe in young people.[102] In Europe a closely related vaccine was developed by Johnson & Johnson subsidiary Janssen in Delft, the Netherlands. Its approval (under the provisional emergency license etc.) was done so hastily that, as Willem Engel of the Dutch anti-lockdown pressure group, *Viruswaarheid,* revealed in March 2021, the accompanying documentation still refers to two doses although this 'vaccine' like the Johnson & Johnson parent company's one is administered in one go.

Johnson & Johnson was not only directly involved in Event 201 but also has connections to the new power bloc where defense and intelligence activities and IT converge. Paul Duprex, a Johnson & Johnson alumnus experienced in gain-of-function research, moved to the University of Pittsburgh Medical Center (UPMC) with support from DARPA. There he worked with the NIH and the Israeli ministry of science and technology on nanobodies, nanoparticles that function as antigens.[103] UPMC is the hub of biodefense academic research in the U.S., for which large sums were spent following the anthrax attacks on Daschle and Leahy (although the anthrax was homegrown). Jeffrey Romoff, its head, wanted to make UPMC the Amazon of healthcare. UPMC was already working on a vaccine against SARS-2 in January 2020; on the basis of the genome of the virus published in China, a vaccine was developed which introduces the spike protein of SARS-2 into the cell by means of an adenoviral adjuvant, i.e. the method used by Johnson & Johnson and AstraZeneca. UPMC, however, went its own way and with a million-dollar grant from CEPI in February, Duprex, together with the Institut Pasteur in France and Austrian Themis (which was acquired by Merck in May 2020), started work on a genetically engineered measles virus vaccine (Merck is a world monopolist in measles vaccines) as the basis for the SARS-2 vaccine. In addition, a second vaccine (in fact the third) was proposed on the basis of anthrax, a very dangerous experiment because in this way an even more risky SARS-2 can arise.[104]

Gilead Sciences. The last company to be mentioned in this context is not in the race for a vaccine but was the first to come up with a drug, *remdesivir*.[105] After the WHO hailed remdesivir as the best bet for a treatment, its share soared, with financial analysts expecting it to generate one-time sales of about $2.5 billion, while adding $12 billion to the company's market value (meanwhile, the WHO has determined that remdesivir works little or not at all).[106] The Wuhan Institute of Virology disputed Gilead's exclusive patent claims for the drug, but that would be a rearguard struggle insofar as it doesn't work. Gilead also developed Tamiflu (marketed by Roche), which reduces the duration of an influenza infection by one day at most and was removed from the drug list by the WHO in 2017.

Gilead Sciences was founded in 1987 by a doctor who was also an alumnus of the Harvard Business School, and as we saw, Donald Rumsfeld had been in charge prior to his appointment to the Pentagon in 2001. George P. Shultz, Reagan's Secretary of State, was also associated with the company, as was Etienne Davignon, former EU Commissioner and Chairman of the Bilderberg Group.[107] Like other drug companies, Gilead had its share of lawsuits, including for alleged delays in the introduction of new antiretroviral drugs to maximize returns on older medications. It has also been criticized for legal tricks that hinder efforts by countries like India to develop generic alternatives.[108] In France, Gilead was in the news after it emerged that the head of the department of infectious diseases at the Saint-Antoine hospital in Paris, who had declared that HCQ was risky, had financial ties with Gilead.[109] Clearly if there is one company that wants to keep HCQ (or Ivermectin) off the market, it would be the maker of remdesivir.

Dilyana Gaytandzhieva, who as we have seen, did such crucial work on American bio-weapon laboratories, also revealed details of a Hepatitis C project run by the CDC with Gilead and the Georgian Ministry of Health in the latter country. This project resulted in at least 249 deaths. Georgian authorities then asked Gilead to help cover up the case. As early as 2016, it was found that in patients who have or have had an infection with the Hepatitis B virus, the antivirals against Hepatitis C can lead to reactivation of the B virus and serious liver problems or even death. In 2020, Gilead donated a new experimental drug for Hepatitis C for use in Georgian patients. Furthermore, Georgia invited Gilead to test remdesivir in the country.[110]

PROTECTION OR INCUBATION OF AN ACTUAL PANDEMIC?

Meanwhile, a large-scale vaccination campaign has started around the globe. A commercial 'Count me in' campaign began circulating on Facebook, in which photo models declare that they want to be vaccinated 'when it is their turn.' And then: 'Goodbye Covid.' Such campaigns, creating a festive atmosphere with song and dance, often crossing the threshold of the outright ridiculous, have proliferated to the far corners of the earth. At the same time, as mentioned, an intensive censorship drive has been mounted by the mainstream media and the IT platforms to keep vaccination warnings away from the public. An urgent open letter to the European Medicine Authority EMA dated 28 February 2021 (cc to the presidents of the European Commission and the EU) with pertinent questions regarding the circulation inside the body of the spike protein and other dangerous effects of the Covid vaccines, signed initially by twelve eminent medical specialists, was not deemed worthy of reply.[111]

The Internet giants, Facebook, Twitter and Google are working closely together to keep this wall of silence effective, as Mark Zuckerberg confirmed in a session of the U.S. Senate Legal Affairs Committee. Zuckerberg also stated there that the task of fact-checking is left to the Poynter Institute and that it operates internationally.[112] Earlier I mentioned that the Poynter Institute is subsidized not only by Facebook but also by the Gates Foundation, eBay owner Pierre Omidyar and George Soros. In the U.S., the Poynter Institute and Facebook have set up the MediaWise Voter Project, which mainly targets young people from minority groups. MediaWise itself refers to a collaboration between Poynter and Stanford University, funded by Google, to teach millions of teenagers to distinguish 'fact from fiction.' The McChrystal Group of ex-JSOC commander Stanley McChrystal and a number of affiliated outfits such as Crosslead are fighting dissidents using the methods of counterinsurgency, isolating leading skeptics as 'health freaks' or bringing them into disrepute otherwise. Crosslead is again affiliated with Kohlberg, Kravis, Roberts (KKR) mentioned in Chapter 3, which has itself branched out into the medical field too. In 2017 KKR acquired WebMD Health Corp, the second largest institutional pharmacy distributor. In a February 2011 article, entitled 'A Prescription for Fear,' *The New York Times Magazine* accused the company of prescribing unnecessary medication.[113]

Meanwhile, after initial denial, the call to limit the freedoms of people who do not want the vaccine is out in the open, no longer dissimulated. Without a vaccine passport, life will become a lot less pleasant. In March 2019, *Scientific American* already gave a taste of what was coming when it advocated excluding those who refuse vaccination from society. The magazine compared such refuseniks to drunken drivers whose behavior endangers the health, safety and livelihood of innocent bystanders.[114]

One cannot but remark the farsightedness of an entire range of institutions and publications which, well before the 'pandemic' was announced, were actively propagating vaccination and the strategy of making freedoms conditional on it. In October 2019 a poll was already circulated at Event 201 showing that 65 percent of the U.S. population would volunteer to be vaccinated against a coronavirus, even if the vaccine was still in the experimental stage. So once the pandemic had been declared, there was a lot to build on. An October 2020 meeting of the WHO's technical advisory group on behavioral influencing discussed the barriers that had to be overcome, especially with regard to the vaccination of children.[115] In particular, the media should be ´educated´ to this end, especially if they report anti-vax feelings. The acceptance of vaccination should be popularized in every possible way. Building trust and preventing people from following rumors and ´conspiracy theories´ about the outsize profits of pharmaceutical companies and the hasty introduction of gene therapies (´info-chemistry´) was a key task.

In January 2021, however, the previously cited survey in 28 countries by PR agency Edelman showed that in 20 of them the willingness to be vaccinated remained below 70 percent. That of course would not be enough to introduce a real Apartheid system given that those who decline are likely in fact to be the better informed, independent-minded section of a population. Even more important was the spectacular collapse of trust in *all* forms of information that has taken place, in particular since the arrival of Donald Trump on the scene, reflected in his incessantly negative mainstream media coverage. A majority of respondents believe that the priority for news organizations is to spread their own views, rather than informing the public. According to Richard Edelman, the rupture is between the 'most informed citizens' (Zelikow's 'relevant political community,' the loyal urban cadre) and the masses. In short, the 'public myth' no longer works. As noted, according to Edelman we have entered the era of information bankruptcy.[116] No small admission from a company that has grown big with 'storytelling' as its PR method.

In the meantime, a unique opportunity has been created in Israel to determine the effect of inoculation with mRNA. The Netanyahu government gave Pfizer a monopoly whilst at the same time exerting unprecedented pressure on the population to get injected. At the time of writing this has been successful in about half of the Israeli people, that is, of the Jewish majority. Palestinian Israelis, on the other hand, are excluded from this 'privilege' as they are second-class citizens, so there is a test population and a control group. Pfizer's CEO, the aforementioned Albert Bourla, stated in an interview that Israel actually functions as a laboratory.[117]

However, the research results from this experiment are not encouraging, in fact, they indicate a possible disaster. In the eight weeks after the start of the vaccination campaign, the total number of Covid deaths doubled compared to the ten months before. Signs are that it is quite possible that those who are vaccinated will become spreaders of the virus and even of mutants that are more dangerous because in a number of cases new symptoms appeared. Like the United Arab Emirates, Great Britain, and Portugal, which have also gone down the path of mass vaccination, a sharp rise in deaths has been recorded. But in Israel, that's only because of Pfizer's gene therapy. In contrast, among Palestinians (as in Jordan), deaths have fallen and in Gaza there is practically no Covid, although this open-air prison camp is one of the most densely populated places on earth.[118]

The WHO has already indicated that vaccination does not necessarily protect against further infection. We saw indeed that the potential for the spread of the virus was not included in the test protocols, neither in Pfizer's nor those of the competition. There may occur something quite different: the normal 'life cycle' of a virus, from highly virulent and dangerous, to more infectious but less dangerous ('virus entropy') may be fundamentally affected or even reversed. In 2015 it was established that chickens which had been vaccinated to protect them against Marek's disease, turned out not to become sick themselves (as expected), but nevertheless were able to pass it on to others in a variant that is much more dangerous (unexpected), something that may currently be the case with the so-called Delta variant affecting vaccinated people.[119] In the Pfizer experiment, the possibility of injecting an ineffective mRNA has been increased by the original requirement that the serum be stored at a temperature of minus 70 to minus 20 degrees Celsius, which was only relaxed in May 2021.[120]

The possibility of a dangerous mutation had also been pointed out by the virologist and vaccine expert formerly employed by GAVI and the Gates Foundation, Dr Geert Vanden Bossche. Vanden Bossche made the news with an open letter warning that 'mass vaccination in a pandemic creates an unbeatable monster.' In this letter, he called for an immediate stop with vaccination or else we may be creating a super-infectious virus that is completely impervious to our most precious weapon against it, the human immune system. After all, we saw that the vaccines use the spike protein similar to that of SARS-2 to bypass the immune system. Like the chickens with Marek's disease, therefore, it can make the recipients of the vaccine asymptomatic spreaders of a potentially deadly variant that carries the risk of extermination of a large part of the human species.[121]

It has been suggested that Vanden Bossche's real aim would be consonant with the interests of his previous employers, GAVI and the Gates Foundation (or CEPI and the vaccine makers for that matter), i.e. to establish a perpetual vaccination regime. However, his call to stop is something else; his warnings moreover are broadly similar to the previously mentioned analyses of Dolores Cahill, Judy Mikovits, Bhakdi *et al.* in their open letter about the dangers of the spike protein, Montagnier in another powerful interview on the *France-Soir* website ('the variants are caused by the vaccines'), etc. Also, Dr Romeo Quijano, a former professor of pharmacology at the University of the Philippines in Manila, has highlighted the dangers of gene editing of viruses which can make the pathogen more aggressive and cause type I interferon reactions that can lead to thrombosis, among other things.[122]

Whatever thought is behind the vaccination campaigns, pursued most radically in the four countries mentioned, with rising mortality as a result, it is clear that we are looking at a completely irresponsible strategy. Assuming that the intention cannot be to make Israeli Jews, Britons, UAE Arabs and Portuguese asymptomatic spreaders of an even more deadly virus, it appears that all of this rests on corruption, thoughtless conformism and the inability to understand the medical consequences of the experimental vaccines. That said, nonetheless, all are consequences of the need to subdue a restless world population, to which the biopolitical complex has added its most revolutionary techniques which anticipate future possibilities, but insufficiently take into account the risks, which may be incalculable. So, what other motives could there be for the mass vaccination campaign?

VACCINATION AS NEUROLOGICAL REMOTE CONTROL

In the same year 2014 in which Rossi and Flagship Pioneering launched Moderna, DARPA's Biological Technologies Office (BTO) launched its In Vivo Nanoplatforms program. It researches implantable nanotechnologies (as a reminder, one nanometer is one billionth of a meter). This concerns, among other things, 'hydrogel,' with which medically relevant values can be transmitted from the cell nucleus via a network such as 5G. The hydrogel is introduced into the body through a dispenser placed under the skin. (Re-) programming of the human organism, 'transhumanism,' a fantasy of the futurologist, Ray Kurzweil, and Google's Eric Schmidt, among others, thus comes within reach. According to the familiar pattern of a publicly funded invention subsequently privatized, the hydrogel technology was transferred to a new company, Profusa. In March 2020 it announced that it had an injectable biochip on offer with which 'Covid could be detected.' The mRNA-serum can be brought into the human body using hydrogel technology in combination with a light sensor technique also funded by DARPA.[123]

Nanotechnology dates back to the late 1950s, when the quantum physicist, Richard Feynman, already expressed his hope that one day it should be possible to arrange atoms one by one according to one's preferences. In the years that followed, important further discoveries were made and in the 1990s a hype ensued that made the nano concept widely known. In the year 2000 Bill

Clinton launched the National Nanotechnology Initiative (NNI). Various universities, NASA, as well as Hewlett-Packard and Motorola had already become involved in the preceding period.[124]

This brings back an aspect already touched on earlier, namely the idea of controlling the human mind, a strategy by far preferable to physical violence as far as the ruling class is concerned. I briefly discussed the MKULTRA program in Chapter 1. In addition to the fake news and misleading information spread by the government and the media, the corruption of universities by making them dependent on project funding, and lowering the level of education generally, there are also (bio) chemical strategies of mind control that are being developed. Viruses too can cause changes in mental activity; as an instance, rabies makes animals infected by the rabies virus extremely aggressive. If infection of the central nervous system with certain viruses occurs, it can lead to irritability, insomnia, hyperactivity and learning difficulties. But there are also beneficial effects; in 2010 it was discovered that the everyday influenza virus can make people more social.[125]

Genetically modified viruses, a technique that is decades old, can be used on both crops and humans. Insects treated with such viruses can then turn crops into genetically manipulated varieties, as well as used as a bioweapon. If humans are infected, this can have serious psychological consequences, ranging from drug addiction to dementia.[126]

SARS-2, which we previously argued is most likely a laboratory-engineered virus, also has neurological effects in many patients, ranging from loss of smell and taste to slowing of brain function, confusion, over-arousal, etc. In SARS-1, MERS and influenza, the numbers for this are much, much lower. The question that arises is whether the creators of the virus in the laboratory where the gain-of-function research was carried out, deliberately aimed at these neurological effects. In all Covid patients who underwent an MRI scan in an intensive care unit at a hospital in Strasbourg, the frontal lobe of the brain was found to be the most affected. The effects of SARS-2 on brain functions turned out to be exceptional: more than half of patients were subject to confusion, and one-third suffered from disorientation and loss of coordination of movement.[127] The frontal lobe in the brain is the part where the specific form of dementia occurs which, among other things, leads to automatic obedience. This was also borne out in Milgram experiments to achieve obedience, which showed a decreased function of this part of the brain, which plays a major role in making moral decisions.[128]

Biotechnology is one of the areas in which nanotechnology can provide major breakthroughs. It would revolutionize medical science because synthetic antibodies in the human body could detect and destroy malignant cells. New vaccines based on nanotechnology will no longer use viruses and bacteria but, for example, small carbon straws that can deliver medication gradually within the body.[129]

However, there is also a clear drawback to this technique. In the Netherlands, DSM, the biotech giant company mentioned above in connection with the selection of AstraZeneca as a vaccine maker, is also active in medical nanotechnology. In addition to using injected nanowires to detect cancer at an early stage, nano-molecules (made of carbon or other materials) can also serve to provide continuous information about particular health conditions and vaccination history, one of the goals of the Gates Foundation and related groups. The DSM research had already been criticized more than a decade ago when the Rathenau Institute in the Netherlands issued a warning that it would make permanent government surveillance possible (the institute called it 'a clandestine population survey').[130]

I already briefly referred to the patent awarded to Microsoft in March 2020 by the World Intellectual Property Organization (WIPO) to use the activities taking place inside of the human body to create cryptocurrency ('data mining'). Such physical activity is comprised of radiation, brain activity, blood and body fluid circulation, organ activity and eye movements, facial and muscle movements and any other activity that can be detected and measured. With that, our bodies and minds can be colonized by Microsoft with the help of new technologies.[131] By means of vaccination, so-called nanosomes, capsules consisting of nanoscale droplets that serve as a transport system to bring a substance to a specific place in the human body, can gain access to it.

The Pfizer and Moderna vaccines include, as noted above, nanoparticles and it turns out these are particles that can produce magnetic fields, which then, like other components of the mRNA vaccine, start to circulate through the body. On social media images of people sticking metal objects to their forehead, upper arm, and other body parts have been shown, initially without a proper explanation, simply as curious side-effects. As it turns out, the nanoparticles that are involved here, are added to the serum in large quantities to facilitate gene delivery into the cell and related processes, a procedure called magnetofection.[132] There is also another, far more spectacular possibility they confer and that is remote-controlling behavior through magnetic stimulation in deep-brain circuits. Chinese research showed it was possible to control the movements of mice in this way.[133]

Vaccination with substances containing magnetic nanoparticles is obviously the ideal form of introducing them into the human body, because it is done under the guise of healthcare, leading people to line up to receive the jab. Ultimately, however, it is about placing humanity under surveillance, even making them react or behave according to instructions from the outside. First of course, each individual must be properly tagged.

BIOMETRIC IDENTIFICATION

In September 2019 the issue of vaccination and identification, the precondition for guaranteed surveillance, was on the agenda of the annual meeting in New York of the ID2020 Alliance. This body was founded in 2016 in the UN building by, among others, Microsoft, the Rockefeller Foundation, and GAVI (the Gates Foundation-affiliated vaccination lobby). There, the Alliance launched a program to introduce a digital identity beacon by means of vaccination—all this in collaboration with the government of Bangladesh, and a number of other partners such as the city of Austin (Texas), the CITRIS Policy Lab at the University of California Berkeley, and the humanitarian organization, Care USA. The plan was to use immunization to create a digital identity, a step that the Bangladesh government representative indicated will ensure access to services and livelihoods. However, since the data mainly end up with American companies and institutions, the U.S. is effectively in control.[134]

The Gates Foundation also funded a similar project at MIT in which a person's vaccination history is applied to the skin via a vaccine using tattoo ink or paint. It relies on nanocrystals (so-called quantum dots) that remain active for five years and emit near-infrared light that can be received by a specially equipped smart phone—all this without the patient's knowledge and invisible to the naked eye. As we saw in the previous chapter, a Chinese visiting researcher was also involved. This beacon makes it possible, says the principal investigator, Kevin McHugh, meanwhile affiliated with Rice University, to quickly and anonymously determine whether someone has been vaccinated.[135] Clearly if only vaccination history was what it was all about, such groundbreaking research would not take place at this level on this scale, because that can also be written in the little booklet that used to be carried by travelers to faraway places.

This transpires also from another project funded by Gates (with MasterCard and again GAVI), 'Trust Stamp.' It involves a biometric digital identity card combined with vaccination data. This was justified with the finding that although 2 to 3 million lives are saved each year through vaccination, data on the immunization status of children is very often incomplete in developing countries (but also in the U.S., Italy or Australia). In 2018, a report on safe travel ('The Known Traveler') was presented at the WEF in Davos, to which Google, Accenture, VISA and Marriott, as well as Interpol and various national police authorities, as well as the Canadian and Dutch governments had contributed. All in order to make travel part of permanent surveillance.[136] Travel, purchases of goods and services, in short, active living—it must all be recorded.

In 2019, a partner deal was also concluded between GAVI, NEC, and Simprints to generate biometric data through vaccination. GAVI chief Seth Berkley stated that this will have to be done in all countries; an ID2020 official did not even mention the health aspect: what it was all about was the need to close an alleged *identity gap*—a prerequisite for control of citizens.[137]

Only with the outbreak of SARS-2 did these projects take off. In April 2020, Gates himself made proposals to use technology to track citizens on a grand scale, urging the U.S. to follow Germany's example. IT applications could then trace where 'we' have been and share the relevant data with institutions interested to know more about the contacts 'we' have had. On this basis, Gates indicated, access to public places and events could be granted again.[138]

The first company to enter this line of business, the UK's VST Enterprises, is working with the UN in the Sustainable Development Goals program, which also includes a series of cyber technology projects that have nothing to do with ecology but are to be implemented in all 193 member states. A new generation of barcode (beyond the square QR code) should prevent it from being tampered with. In addition to agreements with its own government, VST concluded contracts with 15 countries for a Covi pass: the U.S., Canada, Mexico, India, South Africa and a number of EU countries, including the Netherlands (Germany was not mentioned here). This makes it possible for the authorities to request the most sensitive personal and medical information. This violates medical confidentiality, but the technology nevertheless received the 'Seal of Excellence' from the EU. In May 2020, VST signed a contract with the digital health firm and owner of the Covi pass, Circle Pass Enterprises, to integrate the VCode® into the Covi pass. GAVI is a key player in the background here, as it has long recommended a digital health ID for children, along with its ID2020 partners, the Rockefeller Foundation and Microsoft.[139]

Microsoft itself has meanwhile also entered the race for a vax-pass, as have IBM, Oracle and Salesforce, an IT company working for GAVI among others, and where the co-author of the

Rockefeller/GBN Lock Step scenario, Peter Schwartz, is meanwhile vice-president for strategy.[140] In addition, a number of airlines, including Lufthansa and Virgin, have developed their own pass because 'being able to fly again' has turned out a powerful incentive to get the public to accept the hypodermic needle despite lingering skepticism. But then, a real injection is in fact optional after Johns Hopkins University confirmed that mRNA can also be introduced into the nose with a cotton swab disguised as a PCR test.[141]

Microsoft, IT giant Oracle, defense company MITRE, Salesforce, and several smaller companies and medical institutions have joined forces in the Vaccination Credential Initiative (VCI), which aims to create a globally valid vax-pass. These firms are also active in a MITRE-run Covid-19 Healthcare Coalition, which also includes In-Q-Tel (investment arm of the CIA) and data company Palantir. Here, health care turns out very much under the care of the military intelligence apparatus. Google also pops up in this coalition: after all, the company is also active in DNA tests. Sergey Brin's wife, Anne Wojcicki, is CEO of the DNA testing company 23andMe—her sister Susan is CEO of Google subsidiary YouTube, one of the most active executors of corona dissent censorship on social media.[142]

It would seem that the panopticon, the electronic dome prison, is already in operation, run by a small coterie. However, it will still require a whole series of additional innovations to become a reality encompassing the entirety of human society. Nanotechnology is an important route to the future in this respect, but its various applications are still far from being integrated. In the early 21st century, nanotechnology was seen as a 'second creation' because all materials can be rebuilt beginning with their atoms and molecules. For instance, engineers can now incorporate elements that resemble life functions into materials. This would include sensors able to 'see' that a weakening or any other problem occurs and that can recommend taking action.[143] Another example is the aforementioned quantum dots, consisting of several thousand atoms, which can be introduced into the body as medicine dispensers.

Not only can magnetofection induce movement from the outside, but as we saw in the project of Charles Lieber at Harvard with Wuhan University of Technology, nanowires can be integrated into living organisms to register the electrical communication between cells and to gain access to the sensory perception of the vaccinated person (nanowires are so called because they are at least three times as long as thick). Once again, after achieving access by vaccination to the biomass of the 7 to 8 billion individuals who have to be controlled, remote-controlling them is no longer an absurd fantasy. A Swiss specialist in neuro-engineering in this regard warns against manipulation of the human personality by large IT companies. He advocates a 'right to psychological continuity' to avoid personality-altering interventions already, which are being experimented with in the military.[144]

In his study on the subject Joscha Wullweber shows that governments consciously sought to give nanotechnology a positive ring and not make the mistake that has allowed genetic manipulation to be cast in a bad light. Yet the founder of the American computer company, Sun Microsystems, already warned around 2000 about the danger that nanotechnology, genetics and information science would merge and might be able to begin to make self-replicating nanorobots.[145] However, studies on the effects of nanoparticles on algae and water fleas have shown that they increase mortality and that they have an antibacterial effect, which is not always desirable. Once nanoparticles have penetrated the cell (and their uptake in both the cell nucleus and the mitochondrium has been documented—magnetofection is meant to facilitate just that), this can affect cell functions, lead to mutation of the DNA, and so cause what the mRNA vaccines per se are ostensibly *not* meant to bring about: genetic modification of humans. As a result, due to insufficient risk analysis, warns Wullweber, the hard-won acceptance of nanotechnology might be completely undone. The dream of a billion-dollar industry, still largely unregulated, can then turn into a nightmare.[146]

AN ARTIFICIALLY INTELLIGENT HUMANITY?

I n response to the crisis that erupted in the late 1960s and early 1970s, capitalism embarked on a flight forward that has ultimately turned it into a truly world-embracing, integral system. This has projected the subjection of life to market discipline to the farthest corners of the planet. At the same time, the world population has not only spectacularly increased in number but also become restive, especially after the financial collapse of 2008. That is why strategies to restore discipline are being developed whose common element resides in the exploitation of technological breakthroughs in the field of nanotechnology, the IT revolution, 5G, and microbiology, to name only the most important. In principle, there is no reason why life itself, if viewed as the movement of protein-nucleic acid combinations, would not become remotely controllable by artificial intelligence. To achieve this goal, mRNA vaccines with magnetic nanoparticles are an important first step. That should ultimately make it possible to provide global capitalism with a new social-natural basis, perhaps a more durable 'new normal.' And no doubt the oligarchy is in a hurry to achieve just that.

My thesis is that linking human organisms to artificial intelligence for the benefit of the oligarchy that now controls the planet is possible in principle, but in its totality remains wildly premature. It may be possible, but it will need a range of new inventions still to be made and integrated, and also it paradoxically underestimates humanity, which, in order to forestall it, may (and must) become aware that what is being rolled out is intended to prevent a democratic application of the new technologies, which would make possible the self-determination of all.

The technological breakthroughs in the aforementioned fields potentially have consequences that can only be measured on the scale of geophysical eras such as the Holocene, Pleistocene, and so on. On this topic profound insights have been developed in the course of the twentieth century in which the role of information is central. For example, in the 1920s the French theologian, Pierre

Teilhard de Chardin, had already proposed to add to the spatial expansion of the universe a second, chemical-physical *involution*, from simple to increasingly complex, including a dimension of consciousness. That is why, says Teilhard, in every body, starting with the smallest molecule, there is already 'psyche' present in principle. This cannot be perceived as such, any more than a physicist can perceive change of mass at that level, and yet s/he can presume and then calculate it.[147]

We may therefore also assume that 'psyche,' intelligence and information will become all-predominant as an aspect of life. Vladimir Vernadskii, the Russian biologist who developed the concept of the biosphere (the sphere around the earth in which life arises and develops further) in 1924, under the influence of Teilhard enlarged this notion with the concept of *noösphere* (the 'layer' of consciousness, from Greek *nous,* mind). Whereas the biosphere as such implies only evolutionary development, the noösphere adds the dimension of conscious life, implying that mankind as its highest form, can intervene in that development.[148]

This intervention has led us to referring to the Anthropocene (from anthropos, man), the geophysical era in which human life has come to completely dominate the planet. James Lovelock (known for the *Gaia* theory of the self-regulating biosphere, which protects life on earth by keeping the planet cool), even speaks of an era beyond the geophysical Anthropocene, the *Novacene.* In his eponymous book, he speculates about a future in which the cosmos as such comes to consciousness, through the artificial intelligence produced by humanity, which can operate in places where humans cannot survive.[149] This artificial intelligence will be able to support life for hundreds of millions more years, at least if we can prevent a nuclear war and not experience the impact of a large meteorite, because the earth has become too hot for life to return afterwards. So artificial intelligence should not become integrated into military applications that can escape the grasp of humanity.

While the spectacular expansion of the human psyche with artificial intelligence ushers in a new era of limitless possibilities, the original human intelligence remains indispensable (and superior to even the most complete artificial intelligence), because it relies on intuition, the ability to fathom problems based on feeling.[150] Whether this authentic intelligence, which owes its superiority to feeling, moral considerations and intuition, will survive, is not a given. After all, we are already witnessing the efforts of the major IT concerns, in close collaboration with American and other intelligence services, to trace all our movements and thoughts and with their algorithms make deviant 'intelligence' hard to find and, over time, unthinkable.

Under such circumstances the attempt to link mankind to artificial intelligence via nanotechnology in ways intended to perpetuate the power of the ruling order is a dangerous illusion. The authoritarian, if not dictatorial, implications of that project are already evident, although the peculiarity of the Covid crisis, is that the majority of people, incited by media and governments alike, are in fact demanding that 'measures' be taken and even insist on getting vaccinated. And that precisely is the means to establish the link with a system of artificial intelligence operated from the outside.

Contradicting the ideas regarding an artificially intelligent humanity, the associated medical technologies, and the total control that it would hand to those in power around the world today—ideas which are premature in almost all respects—is the notion of a relatively autonomous, potentially self-conscious global population. After all, a breakthrough towards a new, radically democratic society has also come within reach due to the digitization of everyday life. That is what our closing chapter is about.

Endnotes

1. Naomi Klein, *The Shock Doctrine: The Rise of Disaster Capitalism* (Harmondsworth: Penguin, 2007), p. 13.

2. That he actually had in mind neoliberal capitalism as such, as well as the associated market discipline that penetrates deep into the human organism, I will leave aside here. Alexandre Macmillan, 'La biopolitique et le dressage des populations,' *Culture & Conflict,* no. 78, 2010, pp. 39–53.

3. Michel Foucault, *Sécurité, territoire, population: Cours au Collège de France (1977–1978)* [ed. M. Senellart] (Paris: Gallimard-Seuil, 2004), p. 24, emphasis added.

4. Peter C. Gøtzsche, 'Our prescription drugs kill us in large numbers,' *PubMed.gov,* 30 October 2014 (online).

5. Jens Martens and Karolin Seitz, *Philanthropic Power and Development: Who Shapes the Agenda?* (Aachen: Misereor/Brot für die Welt/Global Policy Forum, 2015), pp. 23–24, 28.

6. Cited in Samantha Sault, 'This is what Bill Gates had to say about epidemics, back in 2015,' *World Economic Forum,* 19 March 2020 (online); *FortRuss,* 'Revealed: Dr. Fauci Plotted with Bill Gates Before Pushing COVID Panic and Doubts About Hydroxychloroquine Effectiveness,' 8 April 2020 (online).

7. Joseph A. Camilleri and Jim Falk, *Worlds in Transition: Evolving Governance Across a Stressed Planet* (Cheltenham: Edward Elgar, 2009), p. 427; *Gates Foundation,* 'How we work,' n.d. (online).

8. Flo Osrainik, *Das Corona-Dossier: Unter falscher Flagge gegen Freiheit, Menschenrechte und Demokratie* [Foreword Ullrich Mies] (Neuenkirchen: Rubikon, 2021), pp. 70–71.

9. Leonard G. Horowitz, *Complaint for Injunctive Relief Against Unfair and Deceptive Trade by Civil Conspiracy in Violation of the Florida Whistleblower Act, Civil Rights, and Public Protection Laws.* Civil suit against Pfizer, Inc., Moderna Inc., Hearst Corp. and Henry Schein, Inc., U.S. District Court for Middle Distric of Florida, 1 December 2020 [pdf], pp. 20–22, 25, 48–49.

10. Tim Schwab, 'Bill Gates's Charity Paradox,' *The Nation,* 20 March 2020 (online).

11. Sault, 'This is what Bill Gates had to say about epidemics.'

12. Rosemary Frei, 'Did Bill Gates Just Reveal the Reason Behind the Lock-Downs?,' *OffGuardian,* 4 April 2020 (online).

13. In Martens and Seitz, *Philanthropic Power and Development,* p. 39.

14. Raul Diego, 'Mass-Tracking COVI-PASS Immunity Passports To Be Rolled Out In 15 Countries,' *ZeroHedge,* 29 June 2020 [orig. *MintPress*] (online).

15. Walter Van Rossum, *Meine Pandemie mit Professor Drosten: Vom Tod der Aufklärung unter Laborbedingungen* (Neuenkirchen: Rubikon, 2021), pp. 99, 103.

16. Peter Phillips, *Giants: The Global Power Elite* [foreword W.I. Robinson] (New York: Seven Stories Press, 2018), p. 241.

17. Andrew Edgecliffe-Johnson, 'Trust in governments slides as pandemic drags on: Erosion of early public support around the world threatens vaccine rollouts,' *Financial Times,* 13 January 2021 (online); *Edelman Trust Baromater 2021* [pdf] (online).

18. *Global Preparedness Monitoring Board: A World At Risk. Annual report on global preparedness for health emergencies* (Geneva: World Health Organization and World Bank Group, 2019), p. 6.

19. Walter Van Rossum, *Meine Pandemie mit Professor Drosten. Vom Tod der Aufklärung unter Laborbedingungen* (Neuenkirchen: Rubikon, 2021), pp. 99, 103.

20. Global Preparedness Monitoring Board, *A World At Risk,* p. 30.

21. Horowitz, *Complaint for Injunctive Relief Against Unfair and Deceptive Trade*, p. 35 n. 6.

22. Van Rossum, *Meine Pandemie mit Professor Drosten*, pp. 100–101, 112–13.

23. Global Preparedness Monitoring Board, *A World At Risk*, p. 15.

24. Paul Anthony Taylor, 'European Plans for "Vaccine Passports" Were in Place 20 Months Prior to the Pandemic: Coincidence?,' *Global Research*, 3 April 2021 (online).

25. Osrainik, *Das Corona-Dossier*, pp. 245–46.

26. *European Union, EU/WHO Global Vaccination Summit*, 12 September 2019 (online); a child in the United States at the age of six already has received 55 vaccines. Daan de Wit, *Dossier Mexicaanse griep: Een kleine griep met grote gevolgen* (Rotterdam: Lemniscaat, 2010), p. 81.

27. *World Economic Forum*, '2030Vision—partnering together to scale technologies for the global goals,' n.d. (online).

28. Van Rossum, *Meine Pandemie mit Professor Drosten*, pp. 225–27.

29. Robert F., Kennedy, Jr., 'Before COVID, Gates Planned Social Media Censorship of Vaccine Safety Advocates with Pharma, CDC, Media, China and CIA,' *Global Research*, 12 March 2021 [orig. *Children's Health Defense*] (online); Horowitz, *Complaint for Injunctive Relief Against Unfair and Deceptive Trade*, pp. 2–3; Van Rossum, *Meine Pandemie mit Professor Drosten*, pp. 114–16.

30. Max Parry, 'Is the Global Pandemic a Product of the Elite's Malthusian Agenda and U.S. Biowarfare?,' *Unz Review*, 16 March 2020 (online); Osrainik, *Das Corona-Dossier*, p. 39; Kennedy, Jr., 'Before COVID, Gates Planned Social Media Censorship.'

31. Cited in Kennedy, Jr., 'Before COVID, Gates Planned Social Media Censorship.'

32. Judy Mikovits and Kent Heckenlively, *Plague of Corruption: Restoring Faith in the Promise of Science* [foreword Robert F. Kennedy, Jr.] (New York: Skyhorse, for Children's Health Defense, 2020), pp. 63, 99, 201.

33. Raul Diego, 'An experimental new vaccine developed jointly with the U.S. government claims to be able to change human DNA and could be deployed as early as next year through a DARPA-funded, injectable biochip,' *MintPress*, 17 September 2020 (online).

34. Frank Ryan, *Virolution* (London: HarperCollins, 2009), pp. 140, 190.

35. Mikovits and Heckenlively, *Plague of Corruption*, pp. 143, 152, 155–56; Michael Friedman, 'Metabolic Rift and Human Microbiome,' *Monthly Review*, 70 (3), 2018, p. 70.

36. Horowitz, *Complaint for Injunctive Relief Against Unfair and Deceptive Trade*, pp. 10–11.

37. *Moderna*, 'mRNA Platform: Enabling Drug Discovery & Development,' n.d. (online).

38. Tyler Durden, 'Coronavirus Uses Same Strategy as HIV to Evade, Cripple Immune System: Chinese Study Finds,' *Zero Hedge*, 27 May 2020 [orig. *South China Morning Post*] (online).

39. Horowitz, *Complaint for Injunctive Relief Against Unfair and Deceptive Trade*, p. 29.

40. Whitney Webb, 'Bats, Gene Editing and Bioweapons: Recent Darpa Experiments Raise Concerns Amid Coronavirus Outbreak,' *Unz Review,* 30 January 2020 (online).

41. Michael S. Northcott, email 23 January 2021, in Northcott and Daniel Broudy email exchange on Propaganda and the 'War on Terror' network. January 2021 (online).

42. Ryan, *Virolution*, pp. 178, 191, 198; David Martin and Judy Mikovits cited in Nakia Freeman, 'The mRNA COVID Vaccine Is Not a Vaccine,' *Global Research*, 15 January 2021 (online).

43. Dolores Cahill, 'Why People Will Start Dying a Few Months After the First mRNA "Vaccinations,"' *Wakkeren.Nl,* 3 January 2021 (online); Aoife Gallagher, 'UCD professor asked to resign from EU committee over Covid-19 claims,' *Irish Times*, 13 June 2020 (online).

44. C.-T. Tseng *et al.*, 'Immunization with SARS Coronavirus Vaccines Leads to Pulmonary Immunopathology on Challenge with the SARS Virus,' *PLoS ONE,* 7 (4), 2012 (online). This research incidentally had been motivated by the fear of an intentional outbreak; Greg G. Wolff, 'Influenza vaccination and respiratory virus interference among Department of Defense personnel during the 2017–2018 influenza season,' *Vaccine*, 38 (2), 2020, pp. 350–54 (online).

45. Johnny Vedmore, 'Pfizer's Experimental Covid-19 Vaccine: What You're Not Being Told,' *Unlimited Hangout,* 18 November 2020 (online); David Klooz, *The Covid-19 Conundrum,* n.p. [Apple Books], 2020, p. 30.

46. Fanny Jimenez, 'Tesla und Curevac haben unbemerkt einen Impfstoff-Drucker entwickelt und bereits im Juni 2019 ein Patent dafür angemeldet,' *Business Insider*, 6 September 2020 (online).

47. Martens and Seitz, *Philantropic Power and Development*, p. 35; Osrainik, *Das Corona-Dossier*, p. 205.

48. Webb, 'Bats, Gene Editing and Bioweapons.'

49. Dilyana Gaytandzhieva, 'The Pentagon Bio-weapons,' *Dilyana.Bg*, 29 April 2018 (online).

50. Horowitz, *Complaint for Injunctive Relief Against Unfair and Deceptive Trade,* p. 27; Webb, 'Bats, Gene Editing and Bioweapons.'

51. Pierre Lescaudron, 'Compelling Evidence That SARS-CoV-2 Was Man-Made,' *Sott.net*, 26 June 2020 (online).

52. Julie Beal, 'A Guide to the Ronavax: Understanding the Experimental Coronavirus Vaccines,' *Activist Post*, 27 September 2020 (online).

53. Senta Depuydt, 'Does the coronavirus pandemic serve a global agenda?' *Sott.net* [orig. *Children's Health Defense.org*], 20 March 2020 (online); Klooz, *The Covid-19 Conundrum*, p. 31; Harvey A. Risch, 'Early Outpatient Treatment of Symptomatic, High-Risk Covid-19 Patients that Should Be Ramped up Immediately as Key to the Pandemic Crisis,' *American Journal of Epidemiology,* 189 (11), 2020, pp. 1218–26.

54. 'Accomplished pharma prof thrown in psych hospital after questioning official Covid narrative,' *LifeSite*, 11 December 2020 (online).

55. Philippe Descamps, 'Une médecine sous influence: Des conflits d'intérêts qui suscitent la défiance,' *Le Monde Diplomatique,* November 2020, p. 22.

56. Pierre Barnérias, *Hold Up: Retour sur un Chaos* [documentary], November 2020 (online); Hugues Maillot and Christophe Cornevin, 'Marseille: l'IHU du professeur Raoult perquisitionné lundi,' *Le Figaro,* 14 June 2021 (online).

57. 'Het RIVM schaft hydroxychloroquine aan tegen het coronavirus,' *De Telegraaf,* 28 March 2020; 'Revealed: Dr. Fauci Plotted with Bill Gates Before Pushing COVID Panic and Doubts About Hydroxychloroquine Effectiveness,' *Fort Russ,* 8 April 2020 (online); Depuydt, 'Does the coronavirus pandemic serve a global agenda?'

58. *Newsmax,* 'Breakthrough Drug: Ivermectin Shows "Astounding" Results Against Coronavirus,' 7 January 2021 (online).

59. Mikovits and Heckenlively, *Plague of Corruption,* p. 68.

60. Sammy Almashatet *et al., Twenty-Five Years of Pharmaceutical Industry Criminal and Civil Penalties: 1991 through 2015* (n.p.), Public Citizen, 2016 (online), p. 4.

61. 'Twenty-Seven Years of Pharmaceutical Industry Criminal and Civil Penalties: 1991 Through 2017,' *Public Citizen,* 14 March 2018 (online).

62. Osrainik, *Das Corona-Dossier,* p. 248.

63. 'Pharmaceutical giants fined record €444 million for "abusive practices,"' *The Brussels Times,* 9 September 2020.

64. Descamps, 'Une médecine sous influence,' p. 23; De Wit, *Dossier Mexicaanse griep,* pp. 118–19.

65. Descamps, 'Une médecine sous influence,' p. 23; 'Scandale du Mediator: les laboratoires Servier reconnus coupables de "tromperie aggravée,"' *France24,* 29 March 2021 (online).

66. Horowitz, *Complaint for Injunctive Relief Against Unfair and Deceptive Trade,* p. 11.

67. Martina Groeneveld, 'Corona-vaccins getest op Afrikaanse bevolking: ethisch of niet?,' *Viruswaarheid,* n.d., [2020] (online).

68. Van Rossum, *Meine Pandemie mit Professor Drosten,* p. 122; Klooz, *The Covid-19 Conundrum,* pp. 47–48.

69. Groeneveld, 'Corona-vaccins getest op Afrikaanse bevolking.'

70. Mikovits and Heckenlively, *Plague of Corruption,* p. 112.

71. Descamps, 'Une médecine sous influence,' p. 23.

72. Ibid.

73. Harvey Marcovitch, 'Editors, Publishers, Impact Factors, and Reprint Income,' *PLoS Medicine,* 7 (10), 2010 (online).

74. Jill Ettinger, 'May Cause Outrage: The 9 Biggest Big Pharma Fines Ever,' *Organic Authority,* 1 August 2012 (online).

75. Gerard van Erp, 'Wie is de speciale Coronagezant Feike Sijbesma eigenlijk?,' *Viruswaarheid,* n.d. [2020] (online).

76. 'Mondkapjes, vaccins en tests zijn een aanslag op de schatkist,' *Algemeen Dagblad,* 2 July 2020; 'Rutte geconfronteerd met belangenverstrengeling rond coronavaccin,' *Nine For News,* 21 August 2020 (online).

77. JayTe, 'Red Flags Soar As Big Pharma Will Be Exempt From COVID-19 Vaccine Liability Claims,' *The Duran*, 2 August 2020 (online); 'AstraZeneca to be exempt from coronavirus vaccine liability claims in most countries,' *Reuters* 30 July 2020 (online); Jordan Williams, 'AstraZeneca's no-profit pledge for vaccines has expiration date: Report,' *The Hill*, 8 October 2020 (online).

78. F. William Engdahl, 'What's Not Being Said About the Pfizer Coronavirus Vaccine: "Human Guinea Pigs"?,' *Global Research,* 15 November 2020 (online).

79. *NOS*, 'Europese Commissie wil soepelere regels in jacht op vaccin' [AFP], 29 June 2020 (online).

80. Simon Jack, 'AstraZeneca vaccine—was it really worth it?' *BBC News,* 30 March 2021 (online); *Reuters* dispatches between 1 and 12 December 2020 (online but no longer accessible).

81. Joseph Mercola, 'How COVID-19 Vaccine Trials Are Designed,' *Global Research*, 9 November 2020 (online).

82. Engdahl, 'What's Not Being Said About the Pfizer Coronavirus Vaccine?'; Osrainik, *Das Corona-Dossier*, pp. 220–21.

83. Michel Chossudovsky, 'The GSK-Pfizer Multibillion Dollar Global Vaccine Monopoly: Big Money for Big Pharma,' *Global Research*, 5 January 2021 (online).

84. Jaimy Lee, 'These 23 companies are working on coronavirus treatments or vaccines—here's where things stand,' *MarketWatch*, 6 May 2020; (online); Almashat *et al.*, *Twenty-Five Years of Pharmaceutical Industry*; 2009 figure from *Drugwatch.com* (online).

85. Groeneveld, 'Corona-vaccins getest op Afrikaanse bevolking.'

86. Horowitz, *Complaint for Injunctive Relief Against Unfair and Deceptive Trade*, pp. 9–10, 25.

87. Ibid., pp. 29, 31–33, 38; on Fauci, p. 58.

88. *NPR [National Public Radio]*, 'Bad Optics or Something More? Moderna Executive Stock Sales Raise Concerns,' 4 September 2020 (online).

89. Osrainik, *Das Corona-Dossier*, pp. 203–4; *Wikipedia*, 'Moncef Slaoui.'

90. Engdahl, 'What's Not Being Said About the Pfizer Coronavirus Vaccine.'

91. Peter Doshi, 'Pfizer and Moderna's "95% effective" vaccines—let's be cautious and first see the full data,' *BMJ Opinion,* 26 November 2020 (online); Jan B. Hommel [Jan Bonte], 'Het Pfizer/BioNTech Vaccin tegen het SARS-CoV-2 virus,' *Kritische beschouwingen over de gezondheidszorg*, 21 December 2020 (online).

92. Mercola, 'How COVID-19 Vaccine Trials are Designed.'

93. Jose M. Martin-Moreno *et al.,* 'Covid-19 vaccines: Where are the data?' *BMJ Opinion (British Medical Journal)*, 27 November 2020 (online); Hommel, 'Het Pfizer/BioNTech Vaccin.'

94. 'Le chef du service infectiologie de la Pitié-Salpêtrière, à Paris, commente les dernières données rendues publiques sur deux vaccins,' *Le Parisien,* 8 December 2020 (online).

95. Horowitz, *Complaint for Injunctive Relief Against Unfair and Deceptive Trade*, pp. 2 n.1, 4.

96. Robert Hart, 'WHO Chief Warns Vaccine Won't End Covid-19 Pandemic as Moderna, Pfizer Announce Early Successes,' *Forbes*, 16 November 2020. (online).

97. Cited in Engdahl, 'What's Not Being Said About the Pfizer Coronavirus Vaccine.'

98. *BBC News*, 'Johnson & Johnson to pay $4.7bn damages in talc cancer case,' 13 July 2018 (online)

99. Descamps, 'Une médicine sous influence,' p. 23.

100. Osrainik, *Das Corona-Dossier,* p. 105.

101. *AsiaTimesOnline,* 'Johnson & Johnson pauses Covid Vaccine Trial,' 13 October 2020 (online).

102. Mercola, 'How COVID-19 Vaccine Trials Are Designed'; Doshi, 'Pfizer and Moderna's "95% effective" vaccines.'

103. Whitney Webb, 'Engineering Contagion: UPMC, Corona-Thrax and "The Darkest Winter,"' *Unz Review*, 25 September 2020 (online).

104. Ibid.

105. Jeannie Baumann, 'Hundreds of Corona Patients Allowed to Try Gilead's Ebola Drug,' *Bloomberg Law,* 10 March 2020 (online).

106. *Bloomberg Law*, 'Gilead Surges After WHO Comments on Coronavirus Drug Testing,' 24 February 2020 (online); Osrainik, *Das Corona-Dossier*, p. 226.

107. De Wit, *Dossier Mexicaanse griep*, pp. 42, 108, 120.

108. *Wikipedia,* 'Gilead Sciences.'

109. The person in question defended herself by claiming she was attacked because she was a woman. Descamps, 'Une médicine sous influence,' p. 22.

110. Dilyana Gaytanzhieva, 'Gilead had paid $178 million to doctors to promote drugs despite patient deaths,' *Arms Watch*, 14 October 2020 (online).

111. Sucharit Bhakdi *et al.*, *Urgent Open Letter from Doctors and Scientists to the European Medicines Agency regarding COVID-19 Vaccine Safety Concerns,* 28 February 2021 (online).

112. Horowitz, *Complaint for Injunctive Relief Against Unfair and Deceptive Trade*, pp. 50–51.

113. Ibid., pp 49, 23–24.

114. Cited in Mikovits and Heckenlively, *Plague of Corruption*, p. 160.

115. *WHO Technical Advisory Group on Behavioural Insights and Sciences for Health, Behavioural Considerations for Acceptance and Uiptake of Covid-19 Vaccines* [Meeting Report, 15 October 2020] (Geneva: World Health Organization).

116. *Edelman Trust Barometer 2021*; Edgecliffe-Johnson, 'Trust in governments slides as pandemic drags on.'

117. Gilad Atzmon, 'Pfizer CEO Albert Bourla Admits Israel Is the "World's Lab,"' *Unz Review*, 27 February 2021 (online).

118. Ibid.

119. Mike Whitney, 'Operation Vaxx-All Deplorables: Codename; "Satan's Poker,"' *Unz Review*, 10 March 2021 (online).

120. Vedmore, 'Pfizer's Experimental Covid-19 Vaccine.'

121. Vanden Bosssche cited in Whitney, 'Operation Vaxx-All Deplorables.'

122. Cited in Engdahl, 'What's Not Being Said About the Pfizer Coronavirus Vaccine.' In the Philippines a Sanofi-Pasteur dengue vaccine (of which Fauci and NIAID own the patent and knew of the risks) was given to hundreds of thousands of children causing some 600 deaths until the Manlia government called an end to it.

123. Diego, 'An experimental new vaccine developed jointly'; Freeman, 'The mRNA COVID Vaccine Is Not a Vaccine.'

124. Joscha Wullweber, *Hegemonie, Diskurs und Politische Ökonomie: Das Nanotechnologische Projekt* (Baden-Baden: Nomos, 2010), pp. 156–59, 178.

125. Lescaudron, 'Compelling Evidence that SARS-CoV-2 Was Man-Made.'

126. Gaytandzhieva, 'The Pentagon Bioweapons'; Lescaudron, 'Compelling Evidence that SARS-CoV-2 Was Man-Made.'

127. Julie Helms *et al.*, 'Neurologic Features in Severe SARX-CoV-2 Infection,' *New England Journal of Medicine,* 15 April 2020 [Correspondence] (online).

128. Lescaudron, 'Compelling Evidence that SARS-CoV-2 Was Man-Made.'

129. Wullweber, *Hegemonie, Diskurs und Politische Ökonomie*, p. 162.

130. Mensje Melchior, 'Dokteren met nanotechnologie,' *Medisch Contact*, 2 December 2009 (online).

131. Vandana Shiva, 'Bill Gates' Global Agenda' [excerpt from *Oneness vs. the 1 %*] *Children's Health Defense*, August 2020 (online).

132. Wu Kai *et al.*, 'Magnetic nanoparticles in nanomedicine: A review of recent advances,' *Nanotechnology,* 30 (50), 2019 (online); Bi Qunjie *et al.*, 'Magnetofection: Magic magnetic nanoparticles for efficient gene delivery' [Review], *Chinese Chemical Letters*, 31 (12) 2020, pp. 3041–46 (online).

133. Wu Songfang *et al.*, 'Genetically magnetic control of neural system *via* TRPV4 activation with magnetic nanoparticles,' *Nanotoday*, 39 (preprint, ScienceDirect), 2021 (online).

134. Chris Burt, 'ID2020 and partners launch program to provide digital ID with vaccines,' *BiometricUpdate.com,* 20 September 2019 (online); Osrainik, *Das Corona-Dossier,* pp. 101–3, 109.

135. Kevin J. McHugh, 'Biocompatible near-infrared quantum dots delivered to the skin by microneedle patches record vaccination,' *Science Translational Medicine,* 11 (523), 18 December 2019 (online); *MIT Africa*, 'Storing medical information below the skin's surface,' 3 February 2020 (online).

136. Van Rossum, *Meine Pandemie mit Professor Drosten*, p. 124.

137. Burt, 'ID2020 and partners launch program.' Next to a stern warning against 'conspiracy theories,' the billion-size markets for facial recognition, identification of gestures, etc., are also listed.

138. Cited in Diego, 'Mass-Tracking COVI-PASS Immunity Passports To Be Rolled Out in 15 Countries.'

139. Ibid.

140. Osrainik, *Das Corona-Dossier,* pp. 112–13; *Financieel Dagblad*, 'Amerikaanse techbedrijven storten zich op vaccinatiedata,' 8 February 2021 (online).

141. Jan Walter, 'Johns Hopkins University bevestigt: Vaccinweigeraars kunnen gevaccineerd worden met behulp van PCR test.' *Viruswaarheid*, 11 February 2021 (online).

142. Whitney Webb, 'Silicon Valley and WEF-Backed Foundation Announce Global Initiative for COVID-19 Vaccine Records,' *Unlimited Hangout*, 15 January 2021 (online).

143. Wullweber, *Hegemonie, Diskurs und Politische Ökonomie*, p. 163.

144. Marcello Ienca, 'Do We Have a Right to Mental Privacy and Cognitive Liberty?' *Scientific American*, 30 July 2017 (online).

145. Wullweber, *Hegemonie, Diskurs und Politische Ökonomie*, p. 239

146. Ibid., pp. 246–47, 266, 272.

147. Pierre Teilhard de Chardin, 'The Phenomenon of Man' [1959] in J.J. Clarke, ed., *Nature in Question: An Anthology of Ideas and Arguments* (London: Earthscan, 1993), pp. 183–84.

148. William Rees, 'Scale, complexity and the conundrum of sustainability,' in M. Kenny and J. Meadowcroft, eds., *Planning Sustainability* (London: Routledge, 1999), p. 110.

149. James Lovelock, *Novacene: The Coming Age of Hyperintelligence* [with B. Appleyard] (n.p.: Allen Lane, 2019), pp. 25–27, 37. See also his *The Revenge of Gaia* [foreword C. Tickell] (Harmondsworth: Penguin, 2007).

150. Lovelock, *Novacene*, pp. 19–20.

7.

RADICAL DEMOCRACY
AND DIGITAL PLANNING

In the preceding chapters, we saw that with the Covid crisis, the capitalist ruling class, which has achieved unprecedented levels of concentration and enrichment, is trying to bring a halt to the disparate social protest movements that have sprung up around the world after 2008—until the Covid lockdowns brought them to a standstill. The socio-political crisis that was the result, was therefore not the outcome of a medical emergency; rather, it has been the governments, operating under the auspices of the internationalized state and ultimately, the capitalist oligarchy, which, by their declaration of war on society, have a created a potentially revolutionary situation. What are the possibilities for popular resistance in this unexpected situation, how might it be carried out, and with a view to achieving what ends?

Let us first outline the contours of the kind of alternative society whose breakthrough the ruling classes are attempting to prevent.[1] This breakthrough will necessarily include addressing both how to curtail the power and how to redirect to public benefit the excessive wealth of the oligarchy that has made the achievements of the IT revolution its private property, as well as the socialization of large companies, for which the rise of the passive index funds, discussed in Chapter 3, have paved the way. The Internet is central to the achievements of the IT revolution. All of its applications must be placed at the service of humanity as a whole. This historic task can no longer be avoided because the ruling oligarchy itself and the multi-layered state apparatus it directs, by proclaiming the 'pandemic,' by their subsequent censoring of both expert and popular opposition to its many antisocial policies—effectively an attack on the population—have forced the situation and provoked a showdown. People will therefore have to discover that while there are many routes to liberation from the grip governments have imposed on their societies, there is only one ultimate destination: a society beyond capitalism. This will take different forms in every country, in accordance with the circumstances and with the possibilities existing there, bearing in mind that if there is no going forward, there is no return but to a condition that will be worse.

The orientation needed to change society will not arise from ideological conviction but rather be dictated by the fact that the lockdowns, which in time will cause increased oppression, poverty, disease and hunger, are unsustainable. For perspective on the matter, it's essential to compare this with the two world wars of the 20th century. These wars were partly intended as a response to the sharp increase in social unrest, especially labor unrest, and to the political organization of the working population and the intellectuals associated with it. However, they were followed by social explosions that were even more violent: after WWI the Russian revolution, after WWII, the Chinese, Korean, and Vietnamese revolutions and the decolonization of Asia and Africa, the Cuban Revolution, etcetera.

Although this time the ideological orientation of the movements that precede the lockdowns of 2020–21 is diffuse, even contradictory, a revival in overall unrest is to be expected. According to Klaus Schwab, the oracle of Davos, this is in fact one of the greatest dangers of the lockdowns. He quotes Branko Milanovic, who, writing in *Foreign Affairs* under the title 'The Real Pandemic Danger is Social Collapse,' points out that if governments resort to violence in combating unrest, 'societies could begin to disintegrate.' Schwab even warns that this might well be in full swing already.[2] So while there is no clear ideological impulse, the revolutions of the past were not unambiguous in that respect either. The theoretical self-assurance of the groups that came to power through such revolutions in the long run eventually became an obstacle to further development. Apart from securing food supplies and public order, a revolution (the social reconceptualization and institutionalization of society) must be open-ended if a democratic breakthrough is to continue over time. To cite Slavoj Žižek, the core of the idea of revolution is the willingness to take advantage of the opportunities that present themselves in an uncertain historical situation—without being entirely sure of the outcome, but always with a view to popular wellbeing.[3]

In the corona-related state of emergency that has now been established, it is first of all a matter of reclaiming freedom and fundamental rights, precisely because this time there is no roadmap that prescribes what we have to do 'by the book' (nor is there, on the other hand, an ideology that can be twisted or falsified with a view to protecting the privileges of revolutionaries unwilling to relinquish power). This means that the movement is free to develop across a broad front and that is what a revolution is all about. I start with responding to the question: why is the world-historical IT revolution key to epochal social restructuring?

INDIVIDUAL VERSUS COLLECTIVE: A CONTRADICTION THAT CAN BE OVERCOME?

The socialism of the Industrial Revolution that spawned the workers' movement in the nineteenth century was necessarily collectivist insofar as its organizational hub was in the workplace. But in the contender states such as Germany, Austria-Hungary, Japan after the Meiji Revolution, and later Russia, late industrialization also took place under the auspices of the state; there was no liberal tradition to anchor individualism. The requirements of industrialization and this political deficiency together led to an orientation in which the state played the leading role. After the Russian Revolution and the further formulation and management of its policies from above under Stalin ten years later, this took the form of a state socialism with an authoritarian command economy. Within one generation the Soviet Union metamorphosed into a modern industrial state, which was only possible thanks to a mass psychosis, the whipping up of the population to superhuman performance, and supported by a state terror literally exterminating all individual initiative and autonomy.

As a result, the Soviet bloc, as constituted after the defeat of the Third Reich, eventually lacked the vitality that can only be guaranteed by enabling the free exercise of human creativity and maximum democracy. The command economy in which the center decreed the priorities of production from above began to slow down in the 1960s as a result, first in the most developed socialist republics which had had a democratic tradition, such as Czechoslovakia. The alternatives formulated there ranged from workers' control to the introduction of market mechanisms, but in 1968 Warsaw Pact troops put an end to the movement for renewal of state socialism.[4] Still, the USSR and the Soviet bloc did not collapse until the late 1980s, so world opinion (with further nudging by the Western press) judged the idea of socialism, its problems and possibilities, in the light of the stagnating, authoritarian Soviet example for another twenty years.

At the same time, a new era was dawning in the West, that of the IT revolution. In this evolving environment, individual freedom and dense collectivity come together for the first time in history. In principle, everyone is free to access all desired information and engage with it according to their preferences, and yet at the same time be part of a process of totalizing all choices in a dynamic, highly adaptable process. What used to be separate, enclosed physical spaces, such as a workplace, an administrative or political office or representative assembly, become cybernetically connected into a single grid. As feedback integrates the goings-on in each domain into a unified process, the IT-enabled citizenry enjoys an unprecedented capacity for real-time intervention, from giving an opinion and sharing it, to casting a vote. Necessarily, the resulting social order cannot be fixed on the basis of a previously drawn up blueprint, it is in constant development. This combination of subjective freedom and objective coherence is discernible in every area of activity. Before this potential can be realized, however, this reality must be addressed: the central, interactive network created by the IT revolution, the Internet, is the property of the U.S.-based, capitalist oligarchy. This problematic was already theorized in a passage by Marx in the sketches for *Capital* where he elaborates how, according to the logic of capitalist production, the factors of labor, that is, tools, machines, etc., and the skills on which they are based, step by step are expropriated by the capitalist class. The end point is 'an automatic system of machines ... set in motion by an automaton, a moving force that moves itself. This machine consists of *information*, mechanical and intellectual.' The workers are only links in this automated system; all their knowledge and skill is contained in what Marx calls the *social brain.*[5]

Not a bad prediction if one realizes this was written in the mid-nineteenth century, when the steam engine was still a marvel of engineering; simply by following the logic of the progressive separation of the workers' ownership from their means of existence. We are looking at how human ingenuity and skill merge into the eventual social brain and yet end up being owned by capital—'our' Google, Microsoft, Apple, etc., and the oligarchy behind them. Nevertheless, everything circulating through the Internet, including the Internet of Things, the Internet as a regulating system of production, is ultimately traceable to labor, not just mental but also manual labor.

That is the great contradiction in which our modern relationships are caught. But the Internet is not only owned by capital, read: the big IT corporations, ICANN as the agency that issues the domain names, all concentrated in the United States and covered by the American national security state.[6] It is simultaneously and commonly regarded as a 'public service' for the world population as a whole, which as noted has been able to realize its individuality through this medium. And however distorted by overt and hidden censorship, however disconcerting some of the emanations arising from the darker depths of the 'social brain,' this is the liberating, democratic potential of the IT revolution. To prevent that potential from becoming embedded in reality, the oligarchy and the internationalized state it controls have unleashed a revolution from above. However, the possibilities of the Internet are of far greater significance to the masses, who have never had access to such a powerful tool, than to the powers for whom the IT revolution serves to strengthen their grip on society.[7]

Today practically the entire world population. even in Africa, where there is often a lack of electricity, is connected to the Internet via mobile phone and laptop. Since information, knowledge, of itself is social (one can obtain information without somebody else being deprived of it, unless indeed it is sequestered, as it often has been throughout history), it is only the capitalist regime, by attaching intellectual property rights to information products (scientific journals, medicines, and the like), which prevents universal use in such a manner. Objectively, 'technically,' the new productive forces could enable the world to move forward into a different, more humane type of society; only the forced arrangements of the decaying capitalist society hold it back. Even before this took the form of the overt dictatorship over information flows in the name of a virus, a 'New Deal on Data' was unveiled at the World Economic Forum in Davos in 2009, a year after the financial collapse. The idea was to make the 'suppliers' of data independent owners, entitled to income from their property; the reverse of the appropriation of information by the IT groups described by Shoshana Zuboff. This would boil down to an antitrust drive comparable to the late 19th-century Populist revolt in the United States that actually consolidated the principle of private ownership and broadened its mass base.[8]

However, it is precisely the dogma of 'the market,' the ideological cornerstone of the capitalist regime (its material cornerstone is class power, pure and simple), that is made obsolete by the new IT capabilities, particularly Big Data. The idea of the neoliberal ideologue, Friedrich Hayek, that the mass of information needed to keep a modern economy running is far too great to plan it centrally and that only market relations can keep this complex machinery running, is dated: the reason that planning in the sense of the Soviet-type command economy led to such failures and ultimately, stagnation, was not just the lack of democracy, but the technological resources for its enablement.[9] It is precisely thanks to the cybernetic adjustment enabled by a system in which central planning is linked to digitized individual preferences, that flexible planning can be made a realistic proposition, very much in the way big block corporations respond to customer demand. The same applies to the sphere of government, where likewise policy initiatives can be continuously monitored, interpellated and adjusted.

The fact that 'socialism' is not held in high esteem in our time is based on its persistent conflation with Communism, and the preponderant focus on justified criticism of abuses, and in particular Stalinism as exemplifying the systematic arbitrariness of a state taken over by the political police. The radical Soviet abolition of economic ownership was based on a minimizing of the role that initiative and creativity (freedom) must play in an effort to ensure an equitable distribution of production (in the context of the need to rapidly build modern military industry for the purposes of defense). But 'entrepreneurship' is an indispensable aspect of society that explores and tries out new possibilities and should not be cornered by giant conglomerates converting every newly conceived activity and invention into a monopolistic revenue stream for stock owners. Also, the desire to eradicate religion overall, rather than disempowering those institutionalized forms of it which operated as a wing of the capitalist grip on power, was an unnecessary attack on personal freedom and the role of individual conscience. All these much-bruited shortcomings can be partly traced back to the special circumstances in the Russian empire, where the revolution unexpectedly was thrown back on after the defeat of the projected industrialized world revolutions in 1922–1924—but only partly.

Inspired by the ideological roots from which it was derived, the state-socialist system was nevertheless repeatedly able to reinvent itself (especially under Khrushchev; Gorbachev arrived too late and could only sign the act of capitulation). All aspects of the Soviet experience therefore deserve to be explored in depth, particularly the experiences gained in the USSR (and later in Chile under Allende) with digital planning.

DIGITAL EXPERIMENTS IN SOVIET STATE SOCIALISM

The state socialism of the USSR, then, took shape after the failure of revolutionary efforts elsewhere. In retrospect, this marks the moment when the internal challenge that arose from the industrial revolution—the socialist workers' movement—emerged as an external challenge to the West. As we have seen, this helped to consolidate the ideological unity of the capitalist West until the late 1980s. Accordingly, when state socialism imploded, capitalism too landed in an existential crisis. Yet it took until the turn of the 20th century before this crisis revealed itself in all its intensity and the global elite had to again resort to the politics of fear to bind the population to the existing order.

The command economy instituted under the five-year plans in the late 1920s was intended to transform Soviet society into a modern industrial state within one generation by (initially extreme) coercion, knowing that if that did not work out, the country would again fall prey to imperialism. Without central direction, the USSR would have failed to defeat the Nazi invaders, albeit having done so at unprecedented losses of people and equipment. In the sixties, when growth began to slow after the initial breakneck industrialization, war, and the reconstruction afterwards, a digital transformation presented itself as one of the options that might enable the USSR to achieve a qualitative breakthrough. In a number of ways, what was accomplished back then still matters, even though the project eventually fell victim to a conservative reaction that blocked its revolutionary potential.

Computer design began at the Academy of Sciences in Kiev in the 1940s. Military applications were a priority, and when the

Soviet leadership heard about the automated air defense system being developed in the United States, they wanted a similar system of their own.[10] Ideological barriers that made theories such as cybernetics (flexible adjustment via feedback) suspect were not lifted until after Stalin's death; Party leader Nikita Khrushchev in his speech at the 20th Party Congress in 1956 in which he denounced Stalinism, also advocated steps to introduce factory automation. Plans to make the intended air defense network available for civilian use in peacetime had meanwhile landed on his desk, but there were also immediate blockages. Yet the idea of digitizing the command economy persisted, alongside a school of thought advocating corporate profitability as a lever to improve efficiency.[11]

At the 22nd Party Congress in 1961, Khrushchev again called for the exploitation of digital technologies for use in the planned economy. This period followed the spectacular Soviet space successes such as Sputnik, and the enthusiasm that the USSR was going to overtake the West was now at its peak. Computers were hailed in the Soviet press as 'machines of communism,' and when it came to applying the principles of cybernetics to an entire economy, it was thought that the USSR would be more successful than the Americans. In the U.S., a report for the Council of Foreign Relations established that cybernetics was well-suited to centrally manage a complex society: plan targets could be adjusted without having to wait and see what the results were after the planning period of five years. The CIA warned that the Soviet Union had come a long way in building an integrated information network and that Khrushchev's threat that the Russians would 'bury the West' might become a reality.[12]

In this optimistic climate, the computer center of the Academy of Sciences of Soviet Ukraine began developing an automated economic control system (Russian: *OGAS*), building on the earlier plans for computer-controlled air defense.[13] Alexei Kosygin, then Deputy Chairman of the Council of Ministers, encouraged these activities, but the year in which an elaborate blueprint was finally submitted, 1964, was also the year in which Khrushchev was deposed by his fellow Politburo members and a more cautious line was adopted. Under the conservative Leonid Brezhnev, and with Kosygin as prime minister, the new leadership opted for greater corporate autonomy instead. Thus, they met the concerns of local party and corporate bosses who were not keen on having all the details of their activities recorded by the center's computers.[14] The voice of the workers and the population in general was irrelevant in this stand-off.

TOWARDS A GLOBAL PLAN TECHNOCRACY?

The idea of digital planning now switched to a different track. Kosygin wanted to intensify economic cooperation with the West in order to modernize the Soviet economy; the leadership simultaneously agreed to an American plan for a joint think tank to investigate the problems of a developed industrial society. Kosygin's son-in-law became the top Soviet representative in the newly established International Institute of Applied System Analysis (IIASA) near Vienna and remained in that position until 1986.[15]

This might have become the basis for a global digital planning infrastructure, but the West's take on the IIASA was primarily to use it as an opportunity to undermine the Soviet system; after the neoliberal turn under Thatcher and Reagan, Anglo-American support for the institute effectively ended. Nevertheless, the computer models developed here would be used by the UN and the Club of Rome (set up by Olivetti, Fiat and other corporate pioneers of the East-West trade) to tackle global problems regarding the management of world resources or pollution and would also be suitable for digital planning.

On the Soviet side, Vernadsky's theory of the biosphere of the 1920s, mentioned in the previous chapter, was now revived. Together with American scientists concerned over the casualness with which members of the Reagan administration dismissed the risks of nuclear war, they arrived at the theory of the nuclear winter on the basis of the IIASA computer models.[16] The extinction of life on earth did not necessarily have to be the result of a nuclear war; with the help of complexity theory ('chaos theory') it was concluded that climatological system changes, gradual or through a sudden catastrophe, could also jeopardize human survival.[17] That humankind might one day also be put in deadly danger by its own rulers in the name of preserving public health would likely have been unimaginable at the time.

As noted, the kind of planning involved in a digital system is qualitatively different from planning in a command economy within which a contender state pursues forced industrialization. It is not about simply using computers, but about being able to collect and enter vast amounts of data, Big Data, and then *discovering* outcomes instead of dictating them. The predicament of Soviet planners working with this notion becomes clear when we see that their calculations were passed on to the general public when it came to the dangers of nuclear war and the ensuing nuclear winter, but not when the Soviet economy itself was the subject. When, on the eve of Gorbachev taking office, it turned out that the centrally dictated five percent growth did not emerge from the computer models but only a meager two percent, it did not get past the censorship.[18]

Gorbachev arrived too late to convert the social structures of the command economy into a digital planning system, especially since the earlier market reforms had made the movement towards a restoration of capitalist relations practically irreversible. The efforts to achieve digital planning thus finally broke down, along with the only type of society that had then had the chance of making it a success.[19]

CYBERSYN IN CHILE

A second digital planning experiment took place in Chile under Salvador Allende's Unidad Popular government. Cybernetic control was the starting point here, as Chile had never had a planned economy, so there were no interests associated with previous command economics. The Cybersyn project was based on the idea of an integrated national economy plus cybernetic regulation. Supply problems, strikes, anything that required changes to planned objectives, were recorded with the help of a network of telex machines and then the targets were adjusted accordingly. The project was led by a British computer expert who had been kept out of IIASA on account of his progressive ideas, to protect the institute's presumed non-political profile. In the end, the Cybersyn project, barely implemented, fell victim to Pinochet's bloody coup d'état that ended popular rule.[20]

Needless to say, this did not mark the end of efforts to digitize society, which had been ongoing in the course of corporate production and demand projections, or its use by the capitalist state. As with the Pinochet coup, but on a world scale, the aim of the Covid state of emergency is to thwart a progressive, democratic development exploiting the new possibilities, this time by creating an integral surveillance society. However, by allowing the biopolitical complex to use inoculation as a means of colonizing the human biomass, the oligarchy has become complicit in what can only be described as crimes against humanity.

CRIMES AGAINST HUMANITY, REVOLUTIONARY RECOVERY

The states of emergency that have followed each other since March 2020, and which we have seen were planned from different starting points, have meanwhile thrown entire societies into disarray. This new, comprehensive installment of a fear scenario is nevertheless running out of steam now that the health scare associated with SARS-2 is coming apart. On top of depression and rising suicide rates, notably among the young, as well as economic distress and the health consequences thereof, the presumed way out, the 'vaccines,' have turned out to only compound the deteriorating situation. What is developing before our eyes is that the mass vaccination campaign is creating the predicted disasters in the form of new variants such as Delta, against which the gene therapies are powerless.

Any normal vaccine experiment would have been stopped long ago on the basis of a fraction of the fatalities and serious side effects reported. Not this time, because this is about more important things than public health. In Israel, as we saw the prime laboratory for the Pfizer experiment and the first country to encourage people over 50 to take a third jab as a booster, '78 percent of those 12 and older [are] fully vaccinated, the vast majority with the Pfizer vaccine. Yet the country is now logging one of the world's highest infection rates, with nearly 650 new cases daily per million people. More than half are in fully vaccinated people.'[21] Already in April 2021, three months into the vaccination campaign, it was established that among the age group of 20 to 29, there occurred a 32 percent rise in overall mortality (against the total percentage rise of 22), suggesting that the 'vaccine' is more deadly for the young. This led Paul Craig Roberts, former Assistant Secretary of the U.S. Treasury, to comment that the Israeli government is in fact presiding over a genocide of the Israeli Jewish population, given that, as noted in the last chapter, Palestinians are excluded from this presumed benefit.[22]

Since the aim of the entire Covid scare was never to protect the people's health, but to get them registered on a bio-ID pass system in order to keep them under tighter control, no medical expertise is allowed to halt the inoculation campaign. While a Harvard epidemiologist, Professor Martin Kulldorff, may claim that the case for vax-passes has collapsed, since natural immunity is by far superior to any immunization achieved by way of gene therapy,[23] this will not stop the likes of Macron, Trudeau and other hardliners on the issue to mandate and enforce vax-pass restrictions on people's movement. Quarantine camps in Australia are not set up for medical reasons either, they too are part of the hardening of the clampdown.

Taking all this together, from the proclamation of the pandemic by the WHO in March 2020, and the consequences of the lockdowns and the vaccination campaigns, we are looking at nothing less than premeditated crimes against humanity. Already before the summer of 2021, more than 1,000 lawyers and over 10,000 medical experts, led by the German lawyer, Dr. Reiner Fuellmich, initiated legal proceedings against the CDC, WHO and the World Economic Forum on those grounds. In addition to the incorrectly used PCR tests and fraudulent death certificates claiming Covid as the cause, the application of gene therapies presented as Covid vaccine violates the Geneva Convention of 1949 on the rights of non-combatants in a war situation. Based on the experiences of Nazi medical experiments, the Convention's article 32 prohibits 'mutilation and medical or scientific experiments not required for the medical treatment of a protected person'; whilst article 147 outlaws conducting biological experiments on protected persons as a serious breach of the Convention. In the prior Nuremberg Code of 1947, such experiments had already been designated as crimes against humanity.[24]

Now, irrespective of whether this lawsuit and others like it will result in an actual trial before the International Criminal Court or the European Court of Human Rights, or states' domestic courts, a growing protest movement at the time of this writing is setting limits to the introduction of a digital surveillance society as envisaged by the architects of the 'pandemic.' Certainly, the protesters do not in most cases represent the majority of the population, which remains mentally captive to the state of emergency as well as physically. However, the absurdity of pressing on with the vaccination campaign in spite of growing signs it poses severe health risks, with gene therapies that do not work, all that against a virus that for most poses no risk whatsoever, is rapidly eroding popular acceptance. With every passing day, the discrepancy between the official narrative and the contrary evidence is seeping through the ever-widening cracks in the media firewall. Given that governments, acting for the oligarchy and the internationalized state, will not let go, the question is: will there be a revolutionary response?

As argued in the course of this book, the conditions for a transition to a Big Data democracy and a new ecological socialism, tailored to the circumstances of the 21st century, are here. The aforementioned extreme concentration of ownership, both at the corporate level (IT groups, passive index funds, etc.) and at the level of the oligarchy, essentially simplifies socialization. As in World War I, when the belligerent states established near-complete power over the economy, in the current crisis the conditions for a new society are already being created by the old one in a number of ways.[25] At no point should a prior total collapse of the economy and social life be allowed to happen, as is now being attempted through the Covid state of emergency in order to 'build back better,' the WEF slogan repeated at nauseam by governments the world over. A revolution along those lines is no longer compatible with the sensitivity to disturbance of modern society, as we will unfortunately be discovering soon, if we have not already. Nowadays, a large city cannot do without supplies for half a day, or chaos and looting will erupt.

The transition to a new system will have to produce institutions that operate parallel to the existing ones to the point where they can take over. The Internet demonstrates that the parallel information flow is already in operation. To remove the shackles on it by the IT giants and government institutions (other than carefully calibrated minimum rules of decency and good taste), the Internet must be recognized as a utility and brought into public ownership—as electricity and water, the railways etc. should also be returned to where they have been privatized from. Certainly, the expropriation of the massive IT entities whose rules and objectives are subject to the diktat of the likes of Jeff Bezos, Bill Gates and their colleagues will garner widespread approval. The concrete reorganization of the public ownership structure of the IT infrastructure will have to be thought through, but it will not be a leap in the dark even at this stage.

The Open Data movement, of which Aaron Swartz was an iconic figure (he ended his life when he learned of the penalty waiting for him for publishing academic texts placed behind paywalls), would serve as an example. Open Data wants a data universe parallel to the Big Data of business, 'civic data.' The abundance of data itself already works to generate a culture that moves away from the bourgeois culture of possessive individualism, a myopic individualism that takes no responsibility for the broader questions of human survival. The free availability of this data creates expectations and habits that help build a democratic culture against corporate control and technocratic or otherwise autocratic government.[26] The corollary understanding of regulation would overcome the bureaucratic notion of it, with its inherent tendency to duplicate a social process by a state or quasi-state one. Given cybernetic control that passes through the public domain including representative bodies at all levels, it would boil down to four steps: 1) understanding the desired outcome; 2) measuring, during the ride, whether that result is being achieved; 3) designing algorithms (planning rules intended to make possible adjustment based on new data), and 4) 'Periodic, in-depth analysis of the algorithm's capacity to self-correct and perform as expected.'[27]

The role of the Internet and print and visual media in allowing such in-depth analysis and discussion, part of setting the priorities of social development, obviously require a reorganization of the highly concentrated media ownership as described in Chapter 3. Here, not a transition to public ownership but an anti-monopolistic course with private initiative leading, would appear preferable. Following the expropriation from the current monopolistic owners and a reclamation by their editors and contributors, print and broadcast ventures should be rewarded by subscribers for the quality of their work and allowed to freely branch off and innovate in form and content. Here Patreon might provide an example, billing itself as 'an American membership platform that provides business tools for content creators to run a subscription service. It helps creators and artists earn a monthly income by providing rewards and perks to their subscribers.'[28]

The centralization of financial control into the passive index funds and the largest banks which have followed their lead, likewise prepares the ground for subordinating the economy to society and parliamentary control. Digital planning systems could turn the economy into a *noönomy* in the sense that where possible, the remaining physical effort of work is removed from it so that only the creative tasks remain.[29] Creativity in the digital age becomes the lead sector serving society as a whole. From the creative arts raising the aesthetic quality of life for all, to all forms of self-expression down to DIY, this will raise the level of satisfaction with life as a realm of possibility rather than constraint and compulsion. Art will no longer serve as a mark of class distinction but as a universal human entitlement; culture itself becomes the central characteristic of the social order.[30] State socialism again pioneered certain aspects of this such as mass accessibility to the arts, but it did so in authoritarian ways limiting innovation for fear of losing the link with the untrained eyes and ears.

Education will therefore play a key role in preparing the new generations for the coming new societies, which undoubtedly will be culturally diverse. Because the development in a creative *noönomy* proceeds at such a pace that any 'pre'-training will soon become outdated, a continuous return to the educational sphere is necessary. The French theoretician, Paul Boccara, has written extensively on the idea of a permanent alternation of work and education to keep up with the information society. This would also abolish the labor market, because incomes will be generated by the socialized *noönomy* and redistributed according to need. Boccara's model is therefore different from a basic income, as envisaged in the Great Reset of the WEF, which presumes passivity and without democratic participation, and allows the authorities to switch off access to a digital payment system.[31]

All the above could easily be extended but also might quickly dissipate into unchecked utopian speculation, if we do not ask the question: Who will bring about the actual seizure of power?

In Chapter 2 we saw that roughly speaking, the class structure in the developed part of the world consists of an urban cadre, a sub-proletariat born of immigration likewise concentrated in the big cities, and a partially obsolete, déclassé middle and working class. The majority of the urban cadre are loyal to the existing order which remunerates them handsomely by comparison; and whilst the rest are not well-paid, they are also divided. The urban immigrant population is relatively indifferent to formal politics in the countries where they ended up (at least, first generation immigrants); and rightly or wrongly, the marginalized often feel that their jobs, indeed their world, have been taken from them by the newcomers. As noted, the anti-immigrant nationalist populists see it as their main task to keep this divide alive.

Now in light of glaring inequalities and diminishing life chances, the marginalized former working class and immigrant replacements will have to join a revolutionary movement seeking fundamental change of the social order or perish. It is different for the technical and managerial cadre working for the oligarchy and the system that has produced them. It will depend on how this relatively privileged category will respond to the institution of a digital surveillance society by which governments the world over have countered the unrest among the lower classes and the marginalized populations. As Nikolai Bukharin wrote at the time of the Russian Revolution, giving up their privileged position will always be a laborious process for the cadres, because only in capitalism can this class occupy such a position.[32]

Certainly, in the 1970s the cadre were pushed to the left under the impact of a militant working class, which still had the upper hand, its own organizations and ideology, whilst labor immigration was still in its early stages. For all their obvious limitations, the IIASA, the Club of Rome, and other bulwarks of forward-looking technocracy were part of the progressive drift, as we saw above. However, this alliance, with its roots in the labor scarcity of the late 1960s, still adopted its progressive orientation under the ideological hegemony of the communist and socialist workers' movement and at a further remove, its state-socialist emanations in the east. The critique of really existing socialism never gained a real hold on the cadre who were at best inspired by the idea of a 'march through the institutions' and drifted off into academia as post-modernism and other fashionable but politically useless or counterproductive trends. Hence it was possible for the neoliberal counteroffensive to recruit fractions of the cadre into a privileged role again by attacking the state-socialist and trade union straitjacket, from which, in many cases, they were happy to break free.

This time it is different, because the IT revolution has given cadres as well as the rest of the population the means to actively engage in politics based on information. In this regard, the May 68 movement also produced, amidst the flurry of new ideas challenging socialist orthodoxy, the theoretical foundation for a revolution of a new type.

In his manifesto, *The Society of the Spectacle,* French philosopher Guy Debord wrote that a seizure of power ending the capitalist regime will not be analogous to the way the capitalist bourgeoisie itself came to power.[33] The bourgeoisie emerged as a 'class of the economy,' of economic development; it was a dynamic force, concentrated in the cities and linked to trade other than by royal charter, and early manufacture outside the purview of the guilds. The agricultural economy of late feudalism with its low productivity was no match for this ascendant class. The 'proletariat' in capitalism, on the other hand, in the sense of the class that seeks to abolish class society as such, will not come to power likewise as a 'class of the economy,' surpassing the predatory dynamics of capital by increased productivity. Worker socialism and state socialism as its ultimate historical embodiment have demonstrated that this is impossible. Even the Soviet Union and its bloc, commanding unprecedented natural resources, with a highly developed technical and scientific elite and a superior education system raising the rest of the population to literacy and cultural development, eventually proved unable to overtake the capitalism of the West because creativity, the one field in which it could have done so,[34] in the circumstances remained curtailed.

According to Debord, therefore, the progressive forces can only become superior to the bourgeoisie on the basis of their ability to see beyond the capitalist horizon, as a *class of consciousness.*[35] In the transition from economics to *noönomy,* the very fact that life more and more revolves around information and creativity, unifies all of society. In one of his early writings, Marx argued that one particular class can liberate the whole of society on one condition: that society is in the same situation as that class.[36] When one thinks back on the emancipation of the industrial working class, it will be obvious that this condition was not met—the working class was usually even smaller in numbers and the complex social structure facing it in no way shared its working conditions, living quarters, and health condition, let alone its world view. Today, in spite of widening income inequality nationally and internationally, the whole of society is increasingly organized around the digital infrastructure; everything revolves around the one universal currency, information. The concept of the 1% vs. the 99% has already emerged and resonated. What, then, will be the composition of the 'class of consciousness' cadre, who will take the lead to guide society out of the coming depression caused by the last-ditch capitalist lockdown, assuming a major world war can be avoided?

The Covid myth in this respect will work as the great equalizer on the way to a classless society. *All* parties will discover that the great external shock to cover the oligarchy's seizure of power is a lie. *All* people will face disempowerment, even relative destitution, and long for freedom. Behind this lies the desire for a livable world for everyone, within the community of one's choice and in and in harmony with nature.[37] A world that has freed itself from capital as the all-pervading and all-corrupting social force that also necessarily maintains all those contradictions.

That world is within reach.

Endnotes

1. This chapter builds on a prior version published as 'Democracy, Planning, and Big Data: A Socialism for the Twenty-First Century?,' in *Monthly Review*, 71 (11), 2020, pp. 28–41.

2. Klaus Schwab and Thierry Malleret, *Covid-19: The Great Reset* (Geneva: World Economic Forum, 2020), pp. 83–88. The Milanovic quote is on p. 84; his article in *Foreign Affairs* appeared in March 2020.

3. Cited in Hjalmar Jorge Joffre-Eichhorn, ed., *Lenin 150 (samizdat)* (Hamburg: LockAss Books, 2020), p. 131.

4. Jiří Kosta, *Abriss der sozialökonomischen Entwicklung der Tschechoslowakei 1945–1977* (Frankfurt: Suhrkamp, 1978).

5. Karl Marx, *Grundrisse, Foundations of the Critique of the Political Economy. Rough Draft* [trans. and intro, M. Nicolaus] (Harmondsworth: Penguin, 1973 [1857–58]), pp. 692, 694.

6. Prabir Purkayashta and Rishab Bailey, 'U.S. Control of the Internet: Problems Facing the Movement to International Governance,' *Monthly Review*, 66 (3), 2014, pp. 114, 118–19.

7. Michel Bauwens, Vasilis Kostakis, and Alex Pazaitis, *Peer to Peer: The Commons Manifesto* (London: University of Westminster Press), 2019, pp. 33–34.

8. Timo Daum, *Das Kapital sind wir: Zur Kritik der digitalen Ökonomie* (Hamburg: Nautilus, 2017), pp. 183–84; Shoshana Zuboff, *The Age of Surveillance Capitalism: The Fight for a Human Future at the New Frontier of Power* (London: Profile Books, 2019).

9. Wlodzimierz Brus, *Sozialisierung und politisches System* [trans. E. Werfel] (Frankfurt: Suhrkamp, 1975), pp. 192–93.

10. Benjamin Peters, 'Normalizing Soviet Cybernetics,' *Information & Culture: A Journal of History*, 47 (2), 2012, pp. 169–70, 154.

11. Slava Gerovitch, 'InterNyet: Why the Soviet Union did not build a nationwide computer network,' *History and Technology*, 24 (4), 2008, pp. 338–40; Evsej G. Liberman, *Methoden der Wirtschaftslenkung im Sozialismus. Ein Versuch über die Stimulierung der gesellschaftlichen Produktion* [trans. E. Werfel] (Frankfurt: Suhrkamp, 1974 [1970]), p. 11.

12. Alexander Vucinich, 'Science,' in Allen Kassof, ed., *Prospects for Soviet Society* (New York: Praeger, for the Council on Foreign Relations, 1968), pp. 319–20; Gerovitch, 'InterNyet,' pp. 335–36; Peters, 'Normalizing Soviet Cybernetics,' p. 165.

13. 'Academician Glushkov's "Life Work,"' *Ukrainian Computing*, 2012 (online).

14. Michael A Lebowitz, *The Contradictions of Real Socialism: The Conductor and the Conducted* (New York: Monthly Review Press, 2012), pp. 118–19; Gerovitch, 'InterNyet,' p. 343.

15. Eglė Rindzevičiūtė, *The Power of Systems: How Policy Sciences Opened Up the Cold War World* (Ithaca, NY: Cornell University Press, 2016), pp. 44, 48, 69, & passim.

16. Robert Scheer, *With Enough Shovels: Reagan, Bush, and Nuclear War* (New York: Random House, 1982); John Bellamy Foster, 'Late Soviet Ecology and the Planetary Crisis,' *Monthly Review*, 67 (2), 2015, pp. 9–11.

17. William Rees, 'Scale, complexity and the conundrum of sustainability,' in M. Kenny and J. Meadowcroft, eds., *Planning Sustainability* (London: Routledge, 1999), pp. 109–10; Georgi Golitsyn and Aleksandr Ginzburg, 'Natural analogs of a nuclear catastrophe,' in Y. Velikhov, ed., *The Night After ... Climatic and biological consequences of a nuclear war* [trans. A. Rosenzweig, Y. Taube] (Moskow: Mir Publishers, 1985).

18. Rindzevičiūtė, *The Power of Systems*, p. 146.

19. Manuel Castells, *End of Millennium* Vol. III: *The Information Age: Economy, Society and Culture* (Malden, Mass.: Blackwell, 1998), pp. 47–56.

20. Katharina Loeber, 'Big Data, Algorithmic Regulation, and the History of the Cybersyn Project in Chile, 1971–1973,' *Social Sciences*, 7 (65), 2018

(online); Rindzevičiūtė, *The Power of Systems*, pp. 71–72.

21. Meredith Wadman, 'A grim warning from Israel: Vaccination blunts, but does not defeat Delta,' [*AAAS*] *ScienceMag*, 16 August 2021 (online).

22. Paul Craig Roberts, 'Never Has a Vaccine Harmed So Many,' *Global Research*, 1 September 2021 (online).

23. *FEE* [*Foundation for Economic Education*], 'Harvard Epidemiologist Says the Case for COVID Vaccine Passports Was Just Demolished: New research found that natural immunity offers exponentially more protection than COVID-19 vaccines,' 30 August 2021 (online)

24. Soren Dreier, '1,000 Lawyers and 10,000 Doctors Have Filed a Lawsuit for Violations of the Nuremberg Code' [pdf], *Sista tiden* [translated from the Swedish], 9 May 2021.

25. V. I. Lenin, *The Impending Catastrophe and How to Combat it* [1917], *Collected Works*, vol. 25 (Moscow: Progress, 1972).

26. Eric Gordon and Jessica Baldwin-Philippi, 'Making a Habit Out of Engagement: How the Culture of Open Data Is Reframing Civic Life,' in Goldstein, *Beyond Transparency*, pp. 139–40.

27. Tim O'Reilly, 'Open Data and Algorithmic Regulation,' in Brett Goldstein (ed. with Lauren Dyson), *Beyond Transparency: Open Data and the Future of Civic Innovation* (San Francisco, Calif.: Code for America Press, 2013), pp. 289–90.

28. Patreon bills itself as serving over 200,000 creators. See patreon.com.

29. Sergey Bodrunov, *Noönomy* [English version of the Russian edition], presented at the conference 'Marx in a high technology era: Globalisation, capital and class' (University of Cambridge, 2018), p. 124.

30. Alan Freeman, 'Twilight of the machinocrats: Creative industries, design and the future of human labour,' in Kees van der Pijl, ed., *Handbook of the International Political Economy of Production* (Cheltenham: Edward Elgar, 2015), pp. 370, 374–75.

31. Paul Boccara, *Transformations et crise du capitalisme mondialisé. Quelle alternative?* (Pantin: Le Temps des Cérises, 2008), pp. 31, 37–38; and the original plan in Boccara, *Une sécurité d'emploi ou de formation. Pour une construction révolutionnaire de dépassement contre le chômage* (Pantin: Le Temps des Cérises, 2002).

32. Nicolas Boukharine, *Économique de la periode de transition: Théorie générale des processus de transformation* [trans. E. Zarzycka-Berard, J.-M. Brohm, intro, P. Naville] (Paris: Études et Documentation Internationales, 1976 [1920]), p. 104.

33. Guy Debord, *La société du spectacle* (Paris: Gallimard, 1992 [1967]).

34. Freeman, 'Twilight of the machinocrats.'

35. Debord, *La société du spectacle*, p. 82.

36. Karl Marx, *Zur Kritik der Hegelschen Rechtsphilosophie* [1844], *Marx-Engels Werke*, vol. 1 (Berlin: Dietz, 1972), p. 388.

37. I have developed a non-Eurocentric critique of foreignness in this sense in *Nomads, Empires, States*, volume 1 of *Modes of Foreign Relations and Political Economy* (London: Pluto Press, 2007) (Deutscher Memorial Prize, 2008) and its successor volumes.

INDEX